MINING

AN INTERNATIONAL HISTORY

JOHN TEMPLE

ERNEST BENN LIMITED/LONDON

First published 1972 by Ernest Benn Limited
Bouverie House, Fleet Street, London, EC4A 2DL

© *John Temple 1972*

Distributed in Canada by
The General Publishing Company Limited, Toronto

Book designed by Kenneth Day

Maps drawn by E. A. Chambers

Printed in Great Britain

ISBN 0 510–12960–9

Contents

List of Maps

Introduction

THE IMPORTANCE OF MINING to mankind is obvious for all to see today. Coal and minerals play a decisive role in the economy of modern industrial societies; without them men would be immeasurably poorer. In the last 200 years their production has expanded enormously, an expansion that began in Britain. Until the eighteenth century the mining industry had a history of slow but steady growth. It supplied minerals that were largely used for tools, weapons, and utensils, in the construction of buildings, sewage systems, and aqueducts. Precious metals such as gold and silver were prized for their beauty, their value, and the power they brought. Coal had been used for many centuries by individual craftsmen such as brewers and glass and soap makers but demand was low and increasing only in Britain after the sixteenth century. The mining of the majority of these minerals was limited by the needs of agricultural communities.

From the eighteenth century important developments leading to an enormous expansion of mining resulted in the development of heavily industrialized communities, and emphasized the importance of mining to mankind. The coal industry began to grow rapidly and the discovery that coke could be used to smelt and manufacture iron released the iron industry from its dependence on charcoal. The perfection of the steam engine by James Watt in the 1780s also made the expansion of mining possible because it solved the problem of water control in mines. It also aided underground hauling and raising of coal and ores, and led to the development of the locomotive. Bessemer, Siemens, and others from the mid-nineteenth century onwards, developed processes to manufacture large quantities of steel cheaply, which also led to a further expansion of the coal and iron ore mining industries.

The demand for metals such as copper, lead, and tin rose too, especially in the second half of the nineteenth century. The bulk of these metals, either in their pure form or as alloys, went into the manufacture of machinery and machine tools. Their use in the making of weapons and utensils

5

continued but this was only a minor part of the total product. The growth in the twentieth century of the motor and electrical industries, for instance, has only served to emphasize man's debt to the miner, since mining provides the basic raw materials. The miner in turn owes a debt to the metallurgist, since metals rarely occur in their pure form in nature, and the miner has to depend on the metallurgist to devise processes to refine metals from complex ores which are often poor in quality. The mining of precious minerals such as gold and silver is something of a special case since its expansion was usually unrelated to industrial developments. Men were drawn to the more remote parts of countries such as America, Australia, and South Africa in their search for gold, silver, and diamonds because it brought the prospect of immediate wealth and prosperity. Once they were found, important results followed for the countries concerned.

Within this history of mining two other important themes are explored: the role of the miner in society from ancient times to the present day and the future of the industry in a world where mineral resources are being used up at an alarming rate.

1 Mining up to the collapse of the Roman Empire

SOMEWHERE ABOUT 6000 BC man entered a phase of his development known as the neolithic age, when a new method of living by farming and stock raising replaced a nomadic food gathering existence. This neolithic revolution began in Persia and northern Iraq and spread within the next 1,000 years to the Nile Valley, the coastal areas of Palestine, the upper basin of the river Euphrates, northern Syria, and central Persia. Here people grew their own food, reared their own stock, erected houses and storeplaces, made tools and weapons, wove fabrics, and made pots. In addition they were acquainted with metals but they had not learned how to put their knowledge to proper use. The neolithic revolution spread slowly; 2,000 years elapsed before it reached India and the mainland of Europe and over 3,000 before it was established in Britain. In all these areas the gathering and mining of metals appeared in its wake. Neolithic men became aware of metals but did not at first know how to make use of them.

The earliest evidence we have of their use by man is related to the area known as the Near East. The first metals used were probably copper and gold since both occurred as free metals, that is they were found as simple metals and not as a chemical compound. These metals were used by neolithic men in the Near East as they could be gathered from the surface or from stream beds. One could scarcely describe the collection of pieces of metal lying on the surface as mining, yet it is here that the history of mining must begin.

It is important to realize that although the valleys of the Tigris, Euphrates, and Nile witnessed the first metal-using cultures, metal gathering did not in fact occur here, simply because neither gold nor copper was found within them. Metals were gathered on the surrounding hilly regions – the mountains of the Sinai peninsula, in the north west parts of Asia Minor, and in the Caucasus, for instance; the metal gatherers exchanged their materials for the surplus food that the people of the fertile river valleys produced.

Copper and gold were the first metals gathered in any quantity, the former being especially important. A civilization using considerable quantities of copper was established in Mesopotamia by about 3500 BC and in Egypt by about 3000 BC. From these dates until about 2000 BC western Asia and north east Africa could be described as living in the copper age. Copper was used to make tools and weapons such as axes, adzes, sickles, knives, fish hooks, spear heads, and so on, and societies that were able to master the art of making such goods were able to raise their standard of living since these objects gave them a mastery over their environment and their neighbours which they had not possessed before. Craftsmen used copper to fashion decorative objects such as a copper bull 2 feet in height made in Mesopotamia in about 3000 BC, and a life-size statue of Pepi I of Egypt in about 2300 BC, which are believed to be the oldest metal statues in existence. Gold did not have a practical use; its major appeal was its aesthetic value since its beauty recommended it above all other metals. It was almost as soft as putty and could easily be fashioned into ornaments and trinkets. Even in neolithic graves, flakes and nuggets of gold have been found, suggesting that it was valued as an ornament or a charm. Finely worked golden ornaments were found in graves at Ur in Mesopotamia that belong to the period 2700–2600 BC.

Copper and gold were not the only metals gathered at this time; lead ores occurred in areas where copper was gathered and lead was known in Mesopotamia by about 3500 BC. Silver occurs free in nature but the most plentiful source of silver has been galena, the lead ore which contains a small proportion of the more precious metal. A process whereby silver could be separated from galena was known by 2500 BC. Silver was valued more highly than gold by many ancient civilizations. This was true of the Sumerians and Hittites who used it as the basis of their monetary systems.

From Egypt and Mesopotamia the knowledge of metals was diffused over Europe, first of all to places such as Crete and Troy. The influence of Crete was exerted towards the west, to Sicily, Sardinia, and along the European shores of the Mediterranean as far as Spain, and by 2400 BC copper mining seems to have begun in Almeria in Spain. The influence of Troy was exerted towards the Danube area and between 2200 BC and 2000 BC copper workings were opened up in the Carpathian ranges of Transylvania and Slovakia, in the eastern Alps, in the Balkans, and in the Erzebirge mountains of Bohemia and Saxony. By about 2000 BC, as a result of the establishment of metal working in central Europe, a culture based on copper had been developed over a wide area.

1 Gold vessels of about 2600 BC from the Royal
Graves at Ur in Sumeria, showing the high quality
of the craftsmen's work

2   A copper figure of a bison made about 2300 BC
by Sumerian craftsmen; another example of the fine
quality metalwork of the period

These copper-based cultures of the ancient world were replaced by
cultures using bronze. The date varies from area to area but by 1500 BC
bronze seems to have been used over a large area of Europe and the Near
East. Tin, the vital ingredient that was added to copper to make bronze,
was not widely distributed in Asia or Europe. Bohemia and Saxony were
the areas supplying the largest proportion needed in the ancient world.
The introduction of bronze led to great increases in the quantity and quality
of weapons and tools. For example, the sword was developed from the
dagger and this was possible because bronze was a much tougher material
and could be fashioned into a long and fairly slender blade that did not
break easily.

Iron, although it occurs abundantly in nature, was not successfully
smelted until about 1400 BC. The melting point of iron is 400 °C higher

3   Bronze bucket from an early Iron Age cemetery
dated sixth century BC, showing how metal was
used to make a very fine domestic utensil

than copper and it was the Hittites who solved the problems posed in
making it. There is evidence that iron was costly and rare in early civilizations. In the tomb of Tutankhamen (*c* 1350 BC) there were vast quantities
of precious metals, jewels, and ornaments but it also contained a dagger,
a miniature head-rest, and a bracelet all made of iron. Their presence
within the tomb of a pharaoh of Egypt can only be explained as indicating
the very high value the Egyptians attached to a newly discovered metal.
Hittite methods of iron smelting were not widely known until about 1200
BC and from this date the use of iron spread throughout the Near East.
The techniques for producing it did not penetrate as far as Britain until
about 200 BC and an even longer period elapsed before iron was being
commonly produced in the countries around the Baltic. Iron was a
curiosity in parts of Europe until the classical period of Greece (*c* 500 BC)

and it owed its popularity to the Romans of the republican period (c 100 BC). The ancient peoples of the Near East had used iron for nearly 1,000 years before it replaced bronze as the principal metal of European peoples.

Evidence of mining before the Greek and Roman period is not abundant but it is possible to see some development, It is worth remembering that in many areas the gathering of metal and metal ores that lay on the surface or near the surface continued throughout this period, and in fact well beyond it into medieval and early modern times. Underground mining was carried on over a wide area; two places in particular have yielded much information about techniques, the Nubian Desert in northern Sudan and the Mitterberg in the south Tyrol, where gold and copper were mined respectively.

The earliest gold that came to Egypt was of alluvial origin, that is it was gold washed out from gravels. Alluvial mining precedes vein mining because usually the veins of gold are found by tracing the alluvial gold to its source in the rocks. In addition, the washing of gravel is a much simpler process than the extraction of gold from the rock. It seems that by about 1300 BC underground mining of gold was well established in Nubia and under Egyptian control. There were over 100 mines in this area, the most famous being in the Wadi Hammamet. The earliest mines were nothing but trenches, then galleries or tunnels were driven into the hillside, followed later by inclined shafts. The deepest mine was 292 feet deep and extended for 1,500 feet along a vein. The roof was supported by leaving pillars of rock, but at one mine wooden props were used, an unusual practice in an area where wood was scarce. The miners used oil lamps for lighting the workings and these were fastened to the miners' foreheads. Ore was removed from the mine in panniers. The principal tool was a stone hammer, roughly cuboidal in shape. The oldest map in the world, dated about 1300 BC, is related to mining at Wadi Karein near Hammamet in Nubia. It shows mountains of gold, roads leading to the mines, and the houses of miners. The miners were criminals, slaves, and people who had been exiled to the area by the pharaohs. It seems that whole families could be deported to the mines. Agatharchides of Cnidus, a tutor of one of the Ptolomies, provides a most vivid account of gold mining carried out under Egyptian supervision in the second century BC. The description he gave is quoted almost verbatim by the historian from Sicily, Diodorus Siculus, who visited Egypt about 50 BC. Diodorus described the recruitment of miners:

4   Detail from an Egyptian wall painting of smiths
pouring molten metal into prepared moulds with
others shown holding tongs and blowpipes which
were used as bellows

For the kings of Egypt collect condemned prisoners, prisoners of war, and
others who, beset by false accusations, have been, in a fit of anger, thrown
into prison; these, sometimes alone, sometimes with their entire family, they
send to the gold mines; partly to exact a just vengeance for crimes committed
by the condemned, partly to secure for themselves a big revenue through
their toil.[1]

The use of captive labour in mines was a very common practice in the
ancient world and ill-treatment of miners was equally common as Diodorus
indicates,

. . . they are held constantly at work by day and the whole night long without
any rest, and are sedulously kept from any chance of escape. For their guards

[1] Quoted in T. A. Rickard, *Man and Metals*, Vol. I

are foreign soldiers, all speaking different languages, so the workers are unable either by speech or by friendly entreaty to corrupt those who watch them.[2]

The work seems to have been very exhausting; men, women, and children were employed below the surface in difficult conditions,

. . . they have to get into all sorts of positions and throw to the ground the pieces they detach. And this they do without ceasing to comply with the cruelty and blows of an overseer. The young children make their way through the galleries into the hollowed portions and throw up with great toil the fragments of broken stone, and bring it outdoors to the ground outside the entrance . . . there is no one who seeing these luckless people would not pity them because of the excess of their misery, for there is no forgiveness or relaxation at all for the sick, or the maimed or the old, or for women's weakness, but all with blows are compelled to stick to their labour until worn out they die in their servitude.[3]

The Mitterberg mines of the south Tyrol, worked from about 1600 BC to 400 BC, have yielded little evidence of the miners who worked them but an abundance of material related to mining technique. Mining at Mitterberg was taking place at much the same time as the gold mining just discussed but these miners seem to have been more advanced. They dug not only shafts but adits too. An adit was a tunnel dug from the lowest point in a mine out into a nearby hillside to drain water from mine workings. Efforts were made to ventilate the mines by means of two or more shafts connected by a single level. Access was gained by tree stems with notches cut to form a ladder; similar tree ladders were used until the seventeenth century AD in Europe. Galleries were timbered, some were almost 300 feet in length, and when the walls were soft they were revetted with moss and clay squeezed between boards. Fire setting was used to split the rock, a technique described by Diodorus and widely practised by miners until early modern times. The rock was heated by means of a brushwood fire in order to induce splitting; the process could be accelerated by throwing water on the rock whilst it was hot. Water was brought into the mines by means of hollowed half tree trunks and by buckets. Tools were simple and mostly cast in bronze. Bronze hammers were used to crush the ore in the mine which was graded, sieved, and then raised to the surface in leather bags. Remains of a primitive windlass with three spokes to raise the bags were found at the entrance of one mine. But these mines in

[2] Quoted in T. A. Rickard, *op. cit.*, Vol. I   [3] *op. cit.*, Vol. I

5   Grimes Graves, England. A mining gallery in these Neolithic flint mines, showing the dark outcrop of 'floorstone' flint that was used for axe making

central Europe and Nubia cannot claim to be the first underground mines; this distinction rests with Stone Age people in western Europe.

The earliest mining in western Europe was for flint and it was done by people who lived in the late Palaeolithic and Neolithic Ages. The most famous mines were at Spiennes in Belgium and Grimes Graves in East Anglia in England, and it seems they were especially productive about 2000 BC. The object of all such mining was to obtain nodules of flint of the type most suitable for making tools – axe heads, knives, arrowheads. At Grimes Graves the earliest mines were shallow pits with no galleries but later the miners developed a series of galleries from a central pit. The main shaft from which the galleries radiated was probably as much as 40 feet deep. The miners used flint axes but their principal tools were wedges, picks and rakes of deer antler, and shovels made from the shoulder blades of oxen. Marks found on the sides of the pits fit the picks, many of which

6 Grimes Graves. A deer antler pick, last used by
Neolithic miners 5,000 years ago, lying in one of the
galleries

have been found in the place where their last owner dropped them. We
are fairly certain that the flints were raised to the surface in leather bags or
wicker baskets, probably being hauled by rope over a cross beam of a tree
trunk that spanned the shaft. It is probable that the galleries were lit with
crudely made chalk lamps, which probably burned from a vegetable wick;
chalk lamps have been found and in one series of galleries the soot marks on
the walls can be seen clearly.

The mining operations discussed so far were not extensive but they show
many of the characteristics of mining technique that were to persist
throughout the medieval era. Mines were shallow. Deep mines were the
exception rather than the rule. The method of raising and lowering men
and materials to the surface by a windlass operated by one or two men was
to be widely practised for centuries. Mining tools were primitive, the pick,
shovel, a wedge; the physical strength of the miner was often his most
valuable asset. The only way he could supplement his primitive tools and

brute strength was by fire setting, a technique that had dubious benefits because the smoke caused enormous ventilation problems. Greek and Roman mining does show some advance on the developments discussed so far, both in the scale of organization and in the skill used to devise machines to drain mines.

The Greeks offer the first real evidence we possess of mining on an intensive scale from their silver mines at Laurium, 25 miles from Athens. They were first worked in the sixth century BC but in the fifth century BC they reached their peak. From these mines it is possible to learn something of mining technique and organization. They belonged to the Athenian state, which let out concessions to individuals on the payment of a fee. Open pit and shaft mining were undertaken, although the latter was the most common. Excavations have revealed extensive underground workings. More than 2,000 shafts had been sunk into the ore body, which was honeycombed with a maze of passages. The main shafts were 4 feet by 6 feet in section and some showed evidence of footholds by which the miners climbed them. Some shafts were sloped so as to serve as stairways. The underground passages were only from 2 to 3 feet in height and width, just large enough to admit the miner but clearly too small for effective work. It is obvious that the miner had to dig while on his knees, on his stomach, or on his back. The ore was removed by digging downwards and then passed down chutes to the level below, from where it was carried or dragged in baskets to the shaft. The tools used were simple – a hammer, a chisel, a pick, and a shovel; the metallic parts were made of iron. The roof was supported by pillars of ore; very rarely was timber used. The workings were dry but they were probably badly ventilated because the passages were small and crooked. The main passages were run in duplicate, parallel to each other, and connected by short cross cuts to promote the circulation of air. The mining galleries were lit by oil lamps; each miner was provided with a lamp made of baked clay, having sufficient oil to burn ten hours, which was the length of a shift. The ore was raised to the surface by a windlass or carried up the inclined shafts in baskets made of leather or esparto grass. The scale of Greek mining is shown also by the size of the labour force. This was indicated by Thucydides, who recorded that 20,000 slaves, many of them working in the mines, had fled after the occupation of Decelea by the Spartans in 413 BC. No mining enterprise to this date could show such an enormous concentration of workers.

Mining in the Roman era made two notable advances; the whole of the empire was extensively exploited for its mineral wealth on a scale never seen before, and the problem of mine drainage was tackled in a bold and imaginative way. Roman demands for metal for domestic, military, and agricultural purposes were very considerable and led to an enormous extension of mining in western Europe. The degree of state control over mining varied at different times in the republican and imperial periods but clearly there was never a total state monopoly. Nevertheless, the state was always the largest owner of mines and the expansion of mining owed much to state direction. Lead was needed in quantity for water pipes and drainage systems but the lead ore, galena, was the principal source of silver, and in the Roman world silver was the most regular money of account. This explains why the Romans developed the lead mines of Britain very rapidly after their invasion in AD 43. Copper and tin were made into bronze, an alloy used by the Romans for almost every hard-wearing purpose, and hence deposits of both were extensively exploited. Gold was used in a variety of ways, perhaps none being as important as its use in payment for luxuries – ebony, ivory, pearls and precious stones, spices and perfumes from India, silks and drugs from China.

The most interesting development was the extensive mining of iron ore. The most important iron mine seems to have been at Noricum in Austria, although deposits in Gaul were highly valued too. Spain was valued particularly for its mineral wealth. Pliny the Elder exaggerated somewhat when he claimed that the whole land of Spain abounded in gold, silver, lead, copper, and iron, but clearly he was impressed by its deposits. The scale of mining in Spain was greater than anywhere in the whole empire; in the second century AD the mines of Murcia employed over 40,000 men. Spain seems to have led the industry both in production and technical skill. The Rio Tinto mines had a training school for mining engineers attached to them and trained men were subsequently despatched to responsible positions elsewhere.

The Romans were the first miners to be faced with acute problems of water in mine workings, and they made a unique contribution to mining technique in the drainage devices they developed, namely the water wheel and the Archimedean screw, or cochlea. Wholly successful machines to remove water from mines were not invented until the late eighteenth and early nineteenth centuries, but these devices did much to ease a most difficult problem. By the standards of the time some Roman mines were very deep, for example shafts at El Centenillo in Spain were as much as

7 A Corinthian terracotta vase of the sixth century BC, depicting Greek underground mine workers. One miner loosens ore, a second gathers it in a hod, while a third passes it up the shaft

650 feet deep, the water level was high, and mines had to be drained or abandonment was inevitable. At Cartagena slaves baled out the water using buckets made from esparto grass, made watertight by being soaked in tar, which held about 2 gallons. The rising cost of labour and the increasing depth of the mines condemned this method and in the Rio Tinto mines the water wheel was introduced. It consisted of an open-work wooden wheel about $14\frac{1}{2}$ feet in diameter and it carried 24 buckets or boxes each 15 inches long, 7 inches broad, and 5 inches deep. A slave could work such a wheel with his feet in treadmill fashion. The water was raised 12 feet by each wheel and at Rio Tinto a series of such wheels was placed one above the other in the workings to raise water from a considerable depth.

The Archimedean screw, or cochlea, seems to have been used extensively where the vein of ore was rich. It had long been in use in Egypt to raise water from one level to another before it was used in mines. It consisted of a wooden core, on which were set screw vanes of wood or copper, the whole being cased in a barrel made of planks. It was pivoted by wooden pins on iron beams and when rotated, either by a crank or by a man working treads round the middle of the barrel, it lifted water from one level to the next. Each pump raised water only about 6 feet vertically and obviously a number had to be used in series to remove water from a deep mine. The efficiency of the cochlea can be best assumed from its widespread use, although some historians have doubts concerning its effectiveness. Diodorus Siculus was in no doubt as to its value:

> . . . These screws raise the water by a continuous movement to the outlet of the gallery, drying the bottom of the mine and making it possible to extend the workings comfortably . . . The Egyptian screw is able, with ordinary effort, to throw up, in a marvellous manner, an immense volume of water; and it draws easily from a great depth a stream that it pours forth at the surface of the earth.[4]

Two further points should be noted about these drainage machines. In the first place they were not found in all mines that had drainage problems; for example no trace of such a machine has been discovered in any mine worked during the Roman occupation of Britain. There is every evidence to show that British mines were drained by baling with buckets. These machines were probably fairly expensive to install and expensive in manpower to operate and hence they were only used in the richest mines. Secondly, they were always powered by men; not until the fourteenth century were horses used to work drainage machines.

In the actual technique of mining ore the Romans made one interesting innovation: hushing – the breaking down of softer beds of rock by using a strong current of water. This technique was used in gold mining in north west Spain, central Gaul, and possibly at the Dolaucothy gold mine in Carmarthen. Near Granada in north west Spain they built canals several miles in length to carry water to huge storage tanks and from these the water was discharged on to softer rock. The broken rock could then be sluiced through channels containing a kind of prickly shrub which caught the gold. Strabo tells us that in his day more gold was won by this method than by underground mining. Pliny mentions the same method in connection with tin mining. The use of water either to break up softer rock containing metal ores or to wash ore dug from beneath the surface was to be practised for many centuries after the Romans.

The place of the miner in society throughout this period to the disintegration of the Roman Empire in the fifth century was very inferior indeed. The majority of miners were slaves; the mining industry was one of

---

[4] Quoted in T. A. Rickard, *op. cit.*, Vol. I

8   An excavated section of a Roman shaft in the
Esperanza mine at Rio Tinto, Spain, showing
footholds for climbing

the few throughout the Greek and Roman era that drew its labour almost
entirely from slaves. Very occasionally, miners were well treated, for
example there were pithead baths at one Roman mine in Wales, but usually
the miner was a slave living and working under atrocious conditions. They
frequently lived underground; a mine at Aramo had a vertical entrance,
possibly to prevent workmen escaping. Some Roman miners lived in caves
near mine entrances as remains indicate. Fetters and blocks of stone have
been found at the Rio Tinto mines in Spain showing that miners were
sometimes chained. Inscriptions at Rio Tinto and skeletons from Aramo
prove that longevity was unusual among the miners, and the large number
of skulls in some workings would seem to indicate a high death rate.
Miners in Greek and Egyptian enterprises were no better treated; to be a
miner must have been the lowest occupation even for a slave.

2 The development of mining from the collapse of the Roman Empire to the sixteenth century

THE COLLAPSE OF THE Roman Empire in the west in the fifth century had a serious effect on mining since the empire was the central authority that had governed such activities and its disappearance led to a considerable decline in the output of metal. This decline can be traced to the fourth century when the Roman Empire suffered a series of economic difficulties such as the shrinkage of trade and the decline of towns, but the process was accelerated by the eventual collapse of Roman authority in western Europe. The Germanic invaders who settled in the former Roman provinces contented themselves with ores that lay near the surface, particularly iron that was needed for weapons and farming tools. The decline in mining was a direct result of the unsettled political conditions in Europe, and it was to enjoy only very brief periods of prosperity, such as that during the reign of Charlemagne in the eighth century when conditions were more stable, until the eleventh century. From the fifth to the eleventh century western Europe was menaced by a series of invaders: the Germanic tribes, the Arabs, the Vikings, the Magyars. These peoples generally caused much dislocation to the economic life of Europe, and mining suffered accordingly.

The eleventh century was marked by a mining revival that continued until the fourteenth century; mining was not unique in this respect for the whole of western Europe enjoyed a period of extraordinary social and economic progress. The growth of mineral wealth helps to explain how it was possible to build costly and magnificent cathedrals, abbeys, and churches, and to adorn them with intricate work in glass and metal. The growth of wealth also made possible greater leisure. The contribution of mining to these developments was ignored by almost all contemporary writers except Albertus Magnus. He wrote with great enthusiasm about

9 An early Christian miner of Roman times, from a wall painting in the catacombs. He carries a pick and a lamp, and hammers and drills lie at his feet

the mines of Freiburg and the fine silver they produced. The mining developments were centred largely on the Alpine areas; in the Harz, the Vosges, the Jura, and especially the eastern Alps where gold, silver, lead, copper, and iron were mined. The discovery in 1170 of valuable silver-bearing ores at Freiburg in Saxony led to a rush of miners to the area not unlike the gold rushes of the nineteenth century. Mining in central Europe was dominated by the Germans; mining and colonization of new lands often went hand in hand. Only in coal mining was German influence not felt; in the Low Countries and England this industry developed from the early thirteenth century onwards.

During the thirteenth and fourteenth centuries coal mining evolved from mere grubbing for surface coal to underground mining, although the former was still more common than the latter. The Greeks knew of the existence of coal since they often referred to a series of substances which burned, contained earth, and which no doubt was coal. Such substances were mentioned by Aristotle but the most complete description is given by Theophrastus. The Romans knew something of the qualities of coal but do not appear to have mined it extensively. One of the forts on Hadrian's Wall in Britain contained a quantity of mined coal when it was excavated, indicating that they did use it. However, the first written reference

23

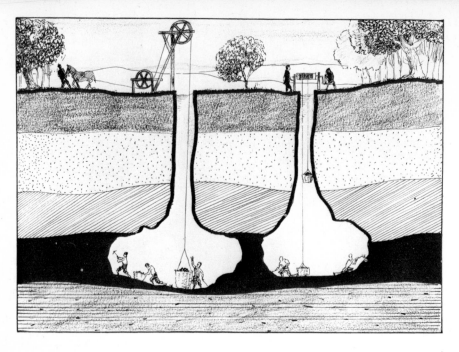

10 An artist's impression of a 'bell pit', showing the primitive tools of the miners and methods of raising coal to the surface in medieval times

to coal mining comes from a charter to the monks of Newbottle Abbey near Preston, dated 1210, giving them the right to mine coal. About the same time coal was mined in Northumberland, since there is a mention of a Sea Coal Lane just outside the walls of London dated 1228. Property in this lane belonged to William of Plessey and it is very probable that he brought coal from Plessey, near Blyth in Northumberland, to London. By the end of the thirteenth century coal was mined in a number of English coalfields, although on a very limited scale in some areas. Coal mining was as hazardous as other types of mining as some references show. For instance, in 1291 a man was killed at Derby when a basket loaded with coal fell on his head, whilst in 1322 Emma, daughter of William Culhare, was killed by choke damp. The second major coal mining area in Europe was the district around Liège in Belgium where coal was mined in considerable quantities in the late twelfth century.

In mining technique few advances had been made since Roman times, and in some respects medieval mining was somewhat primitive compared with Roman mining. There were no mines sunk to a depth of 600 feet as in

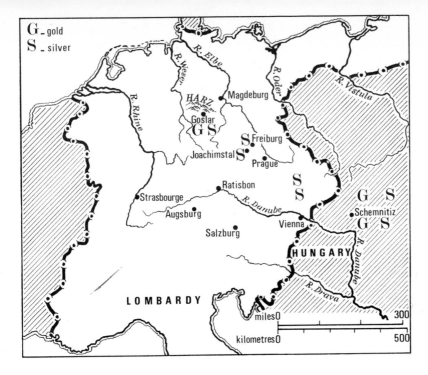

MAP NO. I MEDIEVAL GERMANY – SHOWING PRINCIPAL GOLD AND
SILVER MINING AREAS

Roman times, whilst drainage methods were frequently inferior. Much
of the coal and base ore was mined from open pits or from 'bell pits'. The
latter were so named because they looked like a bell in cross section. A small
shaft was dug through the overlying strata to the coal or metal ore, and when
it was reached the coal or ore was removed round the bottom of the shaft,
moving only a small way under the overlying strata so that the bell shape
was clearly obvious. Rather than risk a sudden roof fall by cutting too far
under the strata, the miners sank a new shaft a few yards away and began
again. From time to time the presence of water in these 'bell pits' must
have led to their abandonment. However, it is very probable that coal and
metal ores were dug from open pits whenever conditions permitted.

Shaft mining was not very common by the end of the thirteenth century
except in the case of the rich silver-bearing ores of central Europe, and
here shafts were usually shallow because miners abandoned them as soon
as water interfered with their work. The short-term answer to this problem
was to remove the water by buckets passed up the shaft by a line of men
or to wind them up the shaft on a hand-turned windlass. The most effective

25

11 An iron miner from the Forest of Dean, with pick, candle, and hod, depicted in a monumental brass of 1463

but expensive method was to dig an adit. In Bohemia in the early four-teenth century experiments were made with adits but some had to be driven more than a mile underground before the workings were reached; a mine had to be very successful to justify this degree of labour and expense. Some efforts were made to construct horse-driven drainage machines to draw water from the mines but with little success.

The mining revival of the eleventh century lasted until the fourteenth and then a slump began which lasted for several generations. This long depression was but one part of the economic troubles of western Europe. Political troubles in the fifteenth century also added to the problems facing mining, notably the Hussite wars (1415–35) that left the Bohemia industry in serious difficulties. From the mid-fifteenth century the mining industry enjoyed a period of prosperity that lasted over a hundred years; partly because of the more settled political conditions and partly because Europe was enjoying a period of prosperity that led to a greater demand for minerals.

Metal mining developed more rapidly because two new and superior techniques were introduced in the separation of metal from ores. About 1451 a certain Johannes Funcken discovered a method of separating silver from silver-bearing copper ores. These ores were plentiful but under-exploited because of the difficulty of extracting the silver. This discovery had an immediate effect on the copper mining industry of central Europe; Saxony, Bohemia, and Hungary became the leading mining areas in Europe. Equally important was the invention of larger furnaces for smelting iron ore. There were three new types, the Stuckofen, the Osmund, and the Catalan. The Stuckofen, discovered in the fourteenth century by the iron workers of the Harz region of western Germany, was the most important and was developed into a blast furnace in the fifteenth century. The Stuckofen furnace could turn out 40 to 50 tons per year, about three times the quantity produced by the normal forge of the period.

We have a fairly comprehensive picture of the most advanced mining and metallurgical techniques about 1550 from the book *De Re Metallica* written by Georg Bauer, who was usually known by his Latinized name of Georgius Agricola. Born in 1494, he settled in 1527 in the small Bohemian town of Joachimsthal which was located in the midst of the most prolific metal mining district of central Europe. He moved to Chemnitz in Saxony in 1533, remaining there until his death in 1555. His book was not published until after his death but it represents the fruits of a lifetime's close observation of mining. It may not provide a representative picture of European mining technique but it gives a rare insight into the most advanced area of the sixteenth century.

One of the most interesting features of mining he described was the use of a variety of mechanical devices powered by water or horses on a scale never seen before. Their principal use was for drainage. Water in the workings, the scourge of the miner in Roman and medieval times, was the most serious of all problems confronting the miner until the introduction of the steam engine in the eighteenth century. The machines described by Agricola were much superior to those used by the Romans. He described no less than six varieties of 'rag and chain' pumps which were so named because the part which carried the water to the surface consisted of a continuous chain to which were attached hollow metal balls. These

12   Three illustrations from Agricola's *De Re Metallica*. *Top left:* a rag and chain pump. Agricola describes six varieties, including one worked by a team of thirty-two horses, in sets of eight at a time. *Top right:* some methods of entering a mine. *Bottom left:* three ventilating devices; those at the top and bottom consist of a series of bellows blowing air into the mine, and the drum rotated by the horse performs the same task

machines were worked by men or horses, but some worked by means of a wheel on the surface which was turned by water. One pump which he describes was driven by a team of thirty-two horses working in groups of eight for four hours, after which they rested for twelve, a further group of eight taking their place. The horses were harnessed in pairs to four projecting arms of an upright axle, and when they rotated it the chain was drawn from the mine lifting a quantity of water in each of the balls. Agricola mentions a mine at Schemnitz in the Carpathian mountains where three such pumps were installed one above the other:

> . . . the lowest lifts water from the lowest sump to the first drain, through which it flows to the second sump; the intermediate one lifts from the second sump to the second drain, from which it flows into the third sump; and the upper one lifts it to the drains of the tunnel through which it flows away. This system of three machines of this kind is turned by ninety-six horses; these horses go down to the machines by an inclined shaft, which slopes and twists like a screw and gradually descends.[1]

The lowest of these machines was 600 feet from the surface but clearly this was an exceptional depth. Deep mining in central Europe was rare, the usual depth was 75 to 80 feet.

Agricola also described drainage machines consisting of nothing more than a windlass which raised water by the bucketful when turned by one or two men. More sophisticated were a series of machines which sucked water from a mine by a series of pistons. Ventilating machinery was also discussed at some length. Some machines consisted of a series of bellows which were operated by hand and drove a current of air round the workings, but the most interesting consisted of a hollow enclosed drum with one fairly small opening on its circumference. When the drum was rotated by a water wheel it forced a current of air along a series of pipes into the mine.

To reach their underground workings the miners used a number of fairly simple methods. The illustration opposite shows them using ladders, being winched down, sliding down an inclined shaft holding a rope, and walking down a series of steps. Their mining tools were crowbars, wedges, picks, hoes, and shovels. Once the ore was loosened it was shovelled into buckets by men described by Agricola as 'shovellers' and winched to the surface. In some of the larger mines, where the workings were some distance from the shaft, the ore was carried along the tunnels in wheelbarrows. Some mines had a series of grooved planks laid along the

[1] G. Agricola, *De Re Metallica*, translated by H. C. and L. H. Hoover

13 This miner using a wheelbarrow is one of the earliest
illustrations we have of wheeled underground haulage

tunnels and along these were pushed small wooden trucks which had a
blunt pin fixed to the bottom which fitted the groove. Agricola claimed that
a truck pushed along such a grooved plank made a noise not unlike the
bark of a dog, hence it was called a 'dog'. More interesting is the fact that
trucks like this anticipated future developments in underground haulage
in the coal mines.

Miners in Agricola's time also used the technique of fire setting used in
classical times. Agricola described fire setting in some detail, showing how
underground fires helped to break up hard rock, but he repeated Pliny's
complaint that the 'foetid vapour' could suffocate miners. Fire setting
was used in some European mines until the late nineteenth century, the
last recorded date being 1885.

One of the most interesting developments in mining was the change in
the miner's status from the convict or slave of classical times to the freeman
in the medieval period, often possessing substantial privileges. Miners at
places such as Freiburg, Goslar, and Joachimstal were exempt from military
service and taxation, largely because these inducements attracted skilled
workers. English tin miners in Devon and Cornwall had the right to prospect anywhere except in churchyards, highways, orchards, and gardens,
paying a tribute of 10 per cent of the produce to the landowner. They had

their own courts under the sole jurisdiction of a warden and no tinner could
be compelled to plead beyond his own court. In addition they had their
own parliaments with the right to legislate for themselves. Lead miners of
Derbyshire, Alston Moor, and the Mendips had their own courts too. Such
freedoms and privileges are a marked contrast to those of the miners of
classical times and represent a recognition of the special skills they possessed.
Their products were valuable, their services were sought after. These
miners were not always employees of some feudal lord, sometimes there
were associations of free miners. In the thirteenth century a German
mining association usually consisted of sixteen men, besides their children
who helped.

Many of the privileges which had helped to raise the status of the miner
in society were to be lost in the fifteenth and sixteenth centuries, largely
because the scale of mining enterprises grew enormously. Such enterprises
of the twelfth and thirteenth century were small and required little capital,

14 A picture from Agricola's
book of a wheel turned by
water from the reservoir (A).
to raise or lower the bucket
(M) containing men and
materials

31

15 Part of a painting by Henri met de Bless (1480–1550) of the exterior of a sixteenth-century copper mine. Notice especially the miner being

but the scale of mining in the fifteenth and sixteenth centuries meant that large sums of money were needed to develop mines, to pay for the drainage machinery, the digging of adits, and sinking of deeper shafts. Mining on this scale was beyond the scope of a self-employed miner or even a group of them. By the middle of the sixteenth century mining companies of 128 shareholders were common in the area where Agricola lived. These shareholders were frequently absentees, they supplied the capital and collected the profits. The direction of such mines was left to managers and foremen who hired hewers, barrowmen, and winders. Thus a clear distinction emerged between the owners of the mine and the workers, resulting in a loss of privileges, since the new owners regarded themselves as heirs to such privileges and chose to ignore them as far as the miners themselves were concerned. This loss of status was most obvious in Germany and led to violent disputes between the mine owners and the miners. A similar loss of privileges happened in England, although somewhat later than in continental Europe. In the seventeenth century many of the independent lead miners of Derbyshire ceased to own the mines they worked, and

winched to the top of the shaft on the right, and
the wide variety of activities at the surface

became wage labourers employed by capitalists. The free tin miners in
Cornwall also became dependent on capitalists. Nevertheless, the miner
was not without a recognized place in society. Agricola was in no doubt
about the miner's status; he devotes Book I of his work to a justification of
mining and to extolling the merits of the miner's occupation.

The prosperity of the gold and silver mines of central Europe came to an
abrupt end when the Spaniards conquered Mexico, Peru, and Bolivia in
the sixteenth century. The discovery of ores extraordinarily rich in silver,
and particularly the opening of the Potosi mines of Bolivia in about 1546,
dealt a serious blow to silver mining. Gold and silver from the New World
could be delivered to Europe more cheaply than the miners of Europe could
produce it. The earliest Spanish invaders of Mexico and Peru took large
quantities of gold and silver from the natives, but this had been gathered
laboriously over a long period of time, almost certainly from river beds
where the metal had been exposed to the action of the water. Large underground reserves were available and these were ruthlessly exploited by the
Spaniards, using natives as miners. In a book published between 1590 and

16 A sixteenth-century engraving by de Bry, depicting
natives of Peru collecting gold from a river bed. The
man on the left is using an ancestor of the 'washing
pan' later used in California

1598 by Jerome Benzoni, called *New History of the New World,* an engraving shows Atahualpa being carried in state and round it there are illustrations of natives mining. The techniques are primitive; the miners are shown using mattocks to dig the ore, panniers for carrying it, and a windlass to hoist it to the surface.

The import of large quantities of gold and silver into Europe had serious effects on the silver mines. Early in the seventeenth century the annual output of silver was perhaps less than a third as great as it had been in the 20s and 30s of the sixteenth century. The collapse of the market for silver mined in Europe brought a reduction in the value of argentiferous copper and lead ores. Conditions were just as bad in other branches of mining. Two hundred years elapsed before mining on the continent of Europe flourished again and by the nineteenth century Britain was dominating the mining industry.

3 Coal mining to the mid-nineteenth century

THE COAL INDUSTRY did not show any signs of rapid expansion until the nineteenth century, except in Britain, where a variety of economic, political, and social factors led to a considerable growth of industry that was not experienced in continental Europe. It is usual to associate British expansion with the enormous development of industry in the late eighteenth and early nineteenth centuries, and the resulting increase in demand for coal. However, it began in the sixteenth century and a lead was established over continental rivals that was not surrendered until the late nineteenth and early twentieth centuries. The annual production of coal until the mid-sixteenth century stood at 200,000 tons per annum; by 1690 the annual output had nearly reached 3 million tons. It is quite clear that there was something like a revolution in the use of fuel in the period from 1550 to 1700. The acute shortage of timber for fuel meant that coal was substituted in industries such as the evaporation of salt water to produce salt, the heating of solutions of alumstone to produce alum, the making of lime, the baking of bricks, and the making of glass. After 1700 the most decisive development of great importance to the coal industry was the discovery of a method of using coal in the manufacture of iron.

The growth of the British coal industry was not accompanied by similar developments in Europe. In France very little expansion occurred until the eighteenth century. It has been estimated that French coal output by the end of the seventeenth century did not amount to more than 50,000 to 75,000 tons per year. In contrast, eight collieries in one manor on the Northumberland and Durham coalfield were yielding over 100,000 tons per year by the eve of the Civil War in 1642. German coalfields were not much further advanced. The chief centres of the industry before 1700 were Saxony, the Ruhr district, and the Wumrevier area round Aachen, but it seems unlikely that the output of all the German fields other than Wumrevier exceeded 150,000 tons at the most at the end of the seventeenth century. It was only the Belgian coalfield that could be compared with the

British fields. Here, the annual output of coal at the end of the seventeenth century was probably no more than a third of that of Durham and Northumberland (approximately 400,000 tons). The Belgian coalfield suffered from the political struggles of the seventeenth century. The collieries of Hainaut province passed from the control of the French back to the control of the Spanish. The entire mining area from Liège to Mons was invaded and ravaged again and again by the armies of Richelieu in their struggle with the Spanish, by the armies of Louis XIV, and by the armies which first the Dutch and then the Continental Alliance raised to attack him. On the other hand, the comparative peace and tranquillity of Britain in this period did much to further the coal mining industry.

The most serious problem facing the coal miners in the seventeenth century was water in the workings, a problem not unique to the coal industry but common to all miners. Most of the pumping devices used in British mines were similar to those described by Agricola in his book. However, these machines had proved inadequate in the deeper mines and strenuous efforts were made to develop a solution to this problem. The first known attempt to use steam power relates to one David Ramsay who obtained in 1631 a patent 'to raise water from low pitts by fire . . . to raise water from low places and mynes, and coal pitts, by a new waie never yet in use'. It should be borne in mind that pumping devices were made for a variety of situations where an efficient machine was urgently needed. They were needed also to drain metal mines and the fens, and to dredge rivers and harbours, but the coal mine owners were the most persistent in trying to find an answer to the problem. To use steam became an English obsession. By 1699 Thomas Savery was demonstrating to the Royal Society an engine

> for raising of water, and occasioning motion to all sorts of mill works, by the impellant force of fire, which will be of great use for draining mines, serving towns with water, and for working all sorts of mills when they have not the benefit of water, nor constant winds.[1]

He published in 1702 a pamphlet entitled *The Miner's Friend* which described a machine designed to raise water from the Cornish tin mines. His engine was very small, however, and it lacked sufficient power to be of much practical value.

Thomas Newcomen, a Cornish engineer, was making investigations with steam-powered machines at the same time as Savery. His work was delayed

---

[1] Quoted in W. H. G. Armytage, *A Social History of Engineering*

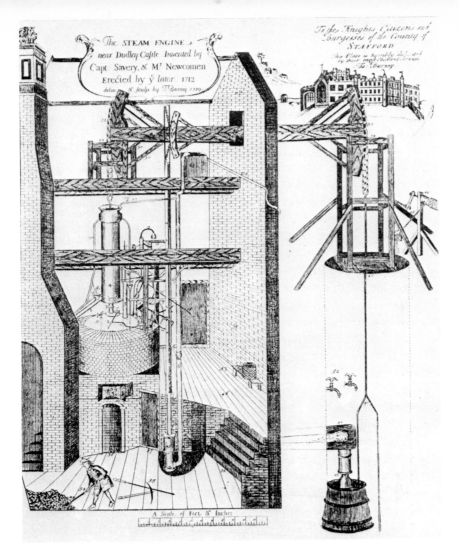

17　A drawing of Newcomen's steam engine installed at a
coal mine at Dudley, Staffordshire, in 1712

by the restrictions of Savery's patent and he did not see his first machine
at work until 1712. He erected his first pumping engine in this year at a
coal mine in Dudley, Staffordshire; it raised 120 gallons a minute over 153
feet. Newcomen's engines were expensive to install but since they were so
efficient they rapidly became an integral part of many productive mines
troubled by drainage problems. It would have taken at least seven of

37

Savery's engines, set one below the other, to lift as much water as a single atmospheric engine standing at the surface. By 1769 there were 120 Newcomen engines in use at coal mines. His engine saved the mines; it was expensive of fuel but this was of little consequence in a coal mining area. Some sixty years after Newcomen's invention James Watt discovered a means to improve it when he succeeded in his project for turning the slow atmospheric engine into an efficient and relatively fast-running engine. This machine was not widely adopted until after Watt's patent expired in 1800 but the engines of Newcomen and Watt had solved the problem that had hindered progress in mining for centuries.

The mid-eighteenth-century expansion of the British coal industry occurred in response to developments in industry, but it also depended upon improvements in mining technique that accompanied these developments. The actual method of mining coal did not see any significant change during this period up to the mid-nineteenth century. The usual method was longwall mining which aimed at removing the whole of a seam in one continuous operation. Coal was cut by a team of hewers working side by side at a long coal face. The working space, running the length of the face, was supported by wooden props, and, as the work advanced further into the seam, the props were moved up and the empty space behind them filled with the waste, the 'goaf' or 'gob'. The coal face was reached from the shaft by roads kept open through the 'goaf'.

This technique was first used in the Shropshire coalfield in the seventeenth century and gradually adopted throughout the country until only the Northumberland and Durham coalfield continued to use the 'pillar and stall' method, which was considered to be more wasteful. This consisted of two operations. The coal was cut up into rectangular blocks or pillars by means of roads driven from the shaft towards the boundary and crossed at right angles by other roads at distances of 60 to 200 yards apart, and known as 'walls' or 'endings'. About 15 to 25 per cent of the coal was worked in this operation. In the second operation, other roads, called 'bords', were driven out from the endings and stalls were opened from these at intervals of 10 to 30 yards apart, and the pillars gradually removed. This second process, called 'robbing the pillar', was not widely practised in the eighteenth century and hence the pillar and stall method was wasteful since large quantities of coal were left behind. The cutting or hewing of the coal required great skill. It was first undercut with a pick. A groove 1 to 2 feet in depth was cut into the lowest part and the mass of coal was sup-

38

18 Digging coal in a very thick seam of a Staffordshire mine using the 'pillar and stall' method. Note the pillars of coal supporting the roof, the wheeled corves on rails, and the use of horses

ported by small wooden props. When this holing process was finished, the props were withdrawn, and the downward pressure of the overlying strata often broke down the coal. If this did not lead to loosening, explosives were used. They were first used in British coal mines in the early eighteenth century, although they were dangerous where there were quantities of explosive gas. Coal cutting was an occupation that was skilful, dangerous, and needed considerable physical strength, and therefore it was done by men.

The haulage of the hewn coal along the galleries or seams to the shaft was done by younger men, sometimes boys, and in some areas, such as Scotland, by women. It was carried in wicker baskets called 'corves' and by the eighteenth century it was generally accepted that corves should hold about 5 hundredweight each. The loaded corves were dragged along the

19 Whim gins like this were frequently used at
mines for winding up coal or metal ores and miners
until the early nineteenth century in Europe

seams to the shaft on a rough sledge called a 'tram'. In the seventeenth
century wheeled trams were introduced into Shropshire pits where wide
seams made this possible. The corves were loaded onto the trams, which ran
on two planks and were prevented from jumping the rails by a guide pin
projecting below the floor of the tram that ran in a groove between the
planks. This was a system very similar to that described by Agricola in his
book. The need for better underground haulage became urgent with the
expansion of coal mining in the eighteenth century. John Curr, a colliery
'viewer' or manager, who worked in mines near Sheffield, invented a
wheeled corve made of wood instead of wicker, which ran on cast-iron
rails. This could be brought to the coal face, filled with coal, pushed along
the gallery to the shaft, raised to the surface, emptied, and all on one set of
wheels. The introduction of cast-iron rails, or 'plates' as they were known,
relieved the barrowmen or coal putters of some of their hardest work.
Their gratitude was recorded by a miner called Thomas Wilson in a poem
written about 1799:

> God bless the man in peace and plenty
> That first invented metal plates;
> Draw out his years to five times twenty,
> Then slide him through the heavenly gates.

In wide seams ponies were used to pull the wheeled trams.

Methods of raising coal to the surface changed rapidly. The earliest miners carried wicker corves on their shoulders up a series of ladders, a practice that persisted in a few Scottish mines until the early nineteenth century. Some mines used a hand-operated windlass fixed over the shaft similar to those described by Agricola. Some seventeenth-century mines began using a device called a 'cog and rung gin' which was a windlass adapted to be worked by horses instead of by manual labour, by the addition of a wheel and pinion arrangement. The horse travelled in a circle round the mouth of the shaft pulling a lever attached to a vertical shaft. The cogs of a horizontal wheel on this shaft worked in the spokes of a small pinion on the windlass or drum shaft, making the latter revolve. When the horse circled the shaft the rope was wound or unwound on the drum, raising or lowering men and materials. The 'cog and rung gin' tended to hinder the work of unloading the corves at the surface and it was replaced in the course of the eighteenth century by the 'whim gin', an improvement since it left the shaft mouth free. The drum round which the rope was wound was removed from the shaft mouth and mounted on a vertical spindle to one side of it. The ropes were led from the shaft to the drum over a series of pulley wheels. When the horses circled the vertical spindle they rotated the drum and the corves were raised to the surface. The miners were also raised from the mine and lowered to the workings in corves. Sometimes they sat astride a wooden beam, sometimes they stood in loops in the rope, or just clung to it as best they could. An engraving dated 1869 shows Polish miners at Wielliczka descending in this way, showing how long this dangerous practice persisted.

Winding the coal from deep mines was a serious problem since horse-driven gins could not generate sufficient power to raise a load of coal to the surface in a single lift. Winding had to be done in two or three stages and this was a very cumbersome process. In some mines it was done by water-powered wheels, a system perfected by Smeaton but still having serious limitations. The problem was not solved until Watt and Boulton developed a steam winding engine in 1784. By the early nineteenth century a number of British mines used these engines and by the mid-century they were in common use and much more powerful than the earlier ones. They could raise any load from the bottom to the top of the deepest mine in one stage.

These developments posed two problems; the need for a device to prevent corves swinging and twisting as they were lifted to the surface, and

20   Polish miners at Wielliczka (1869) descending a shaft in a most dangerous manner – the naked flame of the candle is an invitation to disaster

the need for strong winding ropes to withstand the immense strain imposed by single-stage winding. The first problem was solved by the introduction of wooden and then iron guides built up the sides of the shaft, and the replacement of corves with iron cages into which wheeled tubs were pushed at the mine bottom to be raised more safely to the surface. The introduction of wire ropes in about 1830 solved the second problem. Equally important was the fact that miners could be raised and lowered much more swiftly than before; the iron cage was used for men as well as coal. J. R. Leifchild writing in *The Cornhill Magazine* in 1862 sums up these developments admirably:

> The descent and ascent of these shafts have of late years lost half of their interest, because they have lost all their romance. Now, a vertical pair of 'guides' supplies an upright railway for iron cages, which are not unlike third-class railway carriages on English lines. Into these cages the men creep, and the coal waggons are wheeled. You cannot fall out, nor can the cages fall down; only a carelessly protruding arm or finger may be lopped off. When we were

boys, pitmen either descended in swinging, banging and bounding baskets, or, with true professional dignity, inserting one leg in a loop at the end of the rope and winding their arms around it, 'rode down', defiant of danger and a thousand feet in darkness.

Such bravado shows that Leifchild was not a miner!

Mine ventilation was a serious problem since the deeper the mine the more acute the problem. Adequate ventilation was difficult to secure because of the presence of two gases; chokedamp, a suffocating mixture of nitrogen and carbon dioxide, and firedamp, a mixture of air and methane which exploded readily on contact with a naked light or spark. Ventilation methods varied from coalfield to coalfield. In order to induce a current of air through a mine the most common method was to sink a main shaft and a smaller ventilation shaft, and the natural flow of air between them provided a simple ventilation system. Miners in the seventeenth century adopted the practice of lighting a brazier at the mouth of the ventilation shaft, thus drawing up the foul air. This is first reported to have been used in Staffordshire about 1650 but the method came from Belgium and was described to the Royal Society in 1665. The Belgian system consisted of erecting on the surface, within a few yards of the shaft, a chimney 5 foot square and 28 to 30 feet high. Near the bottom of the chimney an iron cradle was suspended containing burning coals which produced an upward current of air. In mines where there was no firedamp the fire was lit at the bottom of the ventilation shaft, although a complicated system was devised to keep the firedamp from making contact with the fire in mines where the gas was present. As the workings extended and became more complicated, many of the coal faces were inadequately ventilated since the air flowed along the shortest route from one shaft to the other. Doors were built at several points in the mines to guide the flow of air, a system known as 'air coursing' which was first introduced in the Cumberland coalfield.

The next stage in ventilation was the introduction of mechanical devices either to pump fresh air into mines or draw foul air out of them, and in this work engineers in Belgium and France led the way. Until 1840 there was probably not a single mechanical ventilator in regular use in British coalfields, although small primitive fans to blow air into mines were used. In Shropshire miners used a machine called the 'Blow-George'.

... which very much resembled the fan employed by farmers to winnow their corn and may be worked by six men, three and three relieving each other; and in case of the work being continued at night, these six men being relieved by another six. The Blow-George when it is practicable, is worked by a band connected to the steam engine. The air from it is forced into pipes, and sent down with great force to the bottom. This instrument is chiefly used, however. whilst a deep shaft is being sunk, or a level is being carried forward, called a heading, and before the work is sufficiently advanced to be able to make a circuit of air through it.[2]

Belgian and French engineers devised pumps to draw air from the mines. The first air pump was built at Charleroi in 1828 and installed at the mine St Louis in the Mons district in 1830. The subject of mechanical ventilation received a considerable stimulus in Belgium when in 1840 the Academy of Science offered prizes for such machines. Engineers were so successful that by 1850 furnace ventilation had been replaced by mechanical ventilation. In Britain, William Fourness of Leeds was the pioneer in this field for, in about 1835, he invented an exhaust fan and by the mid-1840s it had become one of the recognized systems of ventilation. From the small machines constructed by these men the elaborate methods of modern mine ventilation have been developed.

The dispersal of firedamp was not always achieved by a ventilating machine. During the seventeenth century the practice of 'firing' the gas was used; a writer of the century describes the 'fireman's' work in this way:

> The ordinary way in which the hurt of it [the gas] is prevented, is by a person that enters, before the Workmen, who being covered with wet cloath, when he comes near the Coal-wall, where the Fire is feared, he creepeth on his belly, with a large poll before him, with a lighted candle on the end thereof, with whose flame the Wild-fire meeting, breaketh with violence, and running along the roof goeth out with a noise, at the mouth of the Sink the person that gave fire, having escaped, by creeping on the ground, and keeping his face close to it, till it be overpassed which is in a moment.[3]

The dangers of such a practice were enormous but it persisted until the eighteenth century.

The presence of firedamp made the problem of lighting a mine very serious indeed, since the deeper a mine was sunk the greater the danger of firedamp being present. Naked lights caused many accidents, often of a serious nature involving high loss of life. In the 1730s Carlyle Spedding of Whitehaven devised a 'steel mill' consisting of a steel wheel which was rotated by hand against a slab of flint. The resultant shower of sparks

[2] Quoted in Robert L. Galloway, *Annals of Coal Mining and the Coal Trade*, Vol. I
[3] Sinclair, *A History of Coal Mining* 1672

21 A fireman exploding a pocket of gas in a coal mine

provided a certain amount of light and was believed (though incorrectly) not to ignite gas. A particularly serious accident on the Durham coalfield in May 1812 led to the formation of a society for preventing accidents in coal mines which directed much of its attention to discovering a lamp that could be used safely in mines. A man named Clanny had invented such a lamp somewhere about 1811 or 1812 but it was considered to be too fragile for use. Sir Humphry Davy was invited by the society to offer his ideas and he invented the lamp that bears his name. It consisted of a small oil lamp, the burning wick of which was encased in a cylinder of wire gauze. The air and the gas could both burn but as the gas burned inside the gauze the heat was dispersed by it and the gas in the atmosphere outside did not become hot enough to ignite. The lamp gave a safe light and when the flame changed colour it warned miners of the presence of firedamp, giving them time to evacuate the workings if they felt it necessary. Accidents still occurred because miners persisted in using candles and coal owners thought that the use of the lamp entitled them to send their men to work in places where risk under any circumstances was foolhardy.

In Belgium a similar lamp was devised by Meuseler in 1840. Electric battery-operated lamps did not come into general use until about 1910.

Unless the problems facing coal mining had been largely overcome the rapid development of the industry could not have taken place. The British coal industry led the way in their solution and dwarfed all European rivals except the Belgians, in output and expertise. The area around Mons was the first home of coal mining in Europe. By 1789 there were mines near there over 600 feet deep. The temporary incorporation of the Belgian provinces into France (1797–1815) gave the industry a fresh stimulus. By 1830 the new kingdom of Belgium had about 300 mines and produced 6 million tons a year.

In France the pace of development of the coal industry was much slower for a number of reasons. The French had only two good coalfields; the coalfield of the Departments of the Nord and the Pas de Calais, and the coalfield of the Departments of the Loire and Isère. Their remaining deposits were small and scattered. In the Nord coalfield the Anzin company based on Valenciennes had the tradition of being pioneers, having set up the first mine pump driven by steam in 1732, but in many respects the French coal industry lived under the shadow of Britain and Belgium. In 1852 total French production was less than 5 million tons.

The German states, unlike France, had extensive coal resources but they were not fully exploited before the mid-nineteenth century. The Ruhr field, the deposits near Aachen and the Saar, were only worked effectively after 1815. The Silesian field in the east was not seriously developed until after 1840. The development of the coal industry in Prussia was aided by railway building. The first railway was opened in December 1835 and they were built more quickly in the German states than anywhere else on the continent except in Belgium. In 1850 the German states had over 23,500 miles of railway in operation while France had just under 2,000 miles. The railway system was of considerable benefit to the coal industry which was further helped by the use of new methods of production. Steam pumps, steam-driven winding gear, and new tramways, ideas pioneered in Britain in the late eighteenth and early nineteenth centuries, all helped to expand coal production. The mining laws introduced in Prussia in May 1851 acted as a further stimulus. The first law swept away the strict supervision the state had exercised over private mines. Hitherto it had controlled the extension of new mines, the sale of coal, and the fixing of wages. The second law reduced from 10 per cent to 5 per cent the tax on the gross output of mines. A Prussian law of 1860 abolished the last privileges of the miners' guilds. A miner could now

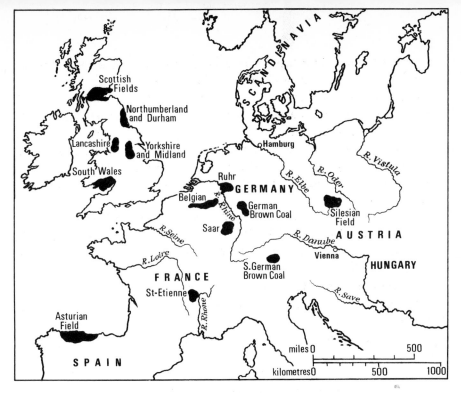

MAP NO. 2   CHIEF EUROPEAN COALFIELDS, MID-NINETEETH CENTURY

move freely from one part of Prussia to another. These three changes in the law were invaluable to the industry. Equally important was the rôle of the state in fostering the development of the economy. The governments of the various German states ran nationalized railways, coal mines, and iron works and in general played a much more decisive role in fostering the economy than the British government. These factors help to explain the rising coal production of the German states by the mid-nineteenth century. Coal production in the three most important fields (Ruhr, Saar, Upper Silesia) was 1 million tons in 1820 and had risen to 6 million by 1850. The emergence of a united Germany by 1871 only accelerated the growth of the coal industry and Britain's supremacy was seriously challenged.

The remarkable rise in coal production to the middle of the nineteenth century can be assessed in a number of ways but the cost in human terms was considerable. Mining life was highly dangerous, accidents were frequent, and the death toll often high. Firedamp had always been one one of the greatest hazards facing miners, but death from roof falls and shaft accidents happened with horrifying regularity. The disaster at the

22 A miner 'holing' or undercutting coal, his only
source of light being a lamp suspended from the roof

New Hartley mine in Northumberland in January 1862 showed the
extraordinary dangers involved. It was worked by only one shaft and
the beam of the pumping engine, weighing more than 40 tons, suddenly fell
into the shaft, trapping underground 204 men and boys. Six days elapsed
before the shaft could be cleared and all the trapped men had died, suffo-
cated by poisonous gases and lack of fresh air. Accidents where large
numbers of miners were killed occurred all too frequently in British and
European mines.

Miners were rarely free agents and able to work for whom they chose
in the early nineteenth century. British miners were hampered by the
yearly bond which bound them to one mine owner, a system that did not
die out until almost the middle of the century in some coalfields. Prussians
wishing to become miners were restricted by the power of the guilds
which hindered the recruitment of more men; these guilds persisted
until 1860. The removal of these restrictions did not occur without a
struggle, yet they did not make mining life any less dangerous or more
attractive.

Perhaps the most distressing feature of mining life was the employment
of children underground, a practice common in Europe and Britain.
Children were frequently employed as trappers, looking after the doors
which guided the ventilating current of air around the workings. From this
they were promoted to become drivers of horses pulling loaded tubs or

48

23   Young boys employed as putters in a British mine
dragging and pushing a loaded wagon underground

wagons along the main underground passages, and eventually they became
putters. The putters were usually young men and their job was to load the
tubs at the face and push them to the main passages. An investigation into
British mines carried out between 1840 and 1842 described conditions
underground for child miners and made them more vivid by a series of
simple drawings to illustrate the findings. The lonely nature of the trapper's
job was shown by the report:

> Their duty consists in sitting in a little hole, scooped out for them in the
> side of the gates behind each door, where they sit with a string in their hands
> attached to the door, and pull it the moment they hear the corves (i.e. the
> carriages for conveying the coal) at hand, and the moment it has passed they
> let the door fall to, which it does of its own weight . . . They have nothing else
> to do . . . they are in the pit the whole time it is worked, frequently about 12
> hours a day. They sit, moreover, in the dark, often with a damp floor to stand
> on, and exposed to drafts . . .
> The ages of these children vary from $5\frac{1}{2}$ to 10 years old . . . There is no hard
> work for these children to do – nothing can be easier; but it is a painful thing to
> contemplate the dull dungeon-like life these little creatures are doomed to
> spend; a life, for the most part, passed in solitude, damp and darkness.[4]

In some Scottish mines children were employed to carry the coal up
ladders in the shaft. Margaret Leveston, six years old, described her
work thus:

[4] Quoted in E. R. Pike, *Human Documents of the Industrial Revolution in Britain*

·24 Women were sometimes employed underground
for heavy manual tasks, as shown in this illustration
of a Belgian mine at Charleroi

> Been down at coal-carrying six weeks; makes ten to fourteen rakes (journeys)
> a day; carries full 56 lbs of coal in a wooden backit (Scots for a shallow wooden
> trough). The work is na guid; it is so very sair. I work with sister Jesse and
> mother; dinna ken the time we gang; it is gai dark.[5]

It is usual to blame the coal owners for allowing this kind of situation in
the mines but it should be remembered that the miners themselves were
guilty of cruelty and indifference to these children as one witness indicated
when giving evidence in 1842:

> There have been cases of maltreatment of children in collieries brought
> before the magistrates . . . The maltreatment was always according to bar-
> barous rules among the workers themselves, inflicting punishment on
> supposed delinquents, generally by holding the head fast between the legs of
> another, and inflicting each a certain number of blows on the bare posteriors
> with pieces of wood, called 'cuts', about a foot long and an inch in diameter,
> used as tokens to distinguish one man's tubs from another. However the one
> punished may cry, they stick to him; and in the last case, where a hungry lad
> had stolen a pit dinner, they mangled his body seriously.[6]

An Act was passed in 1842 forbidding the employment underground of
children under ten and women. Child employment under equally harsh
conditions was common in many parts of Europe and laws to prevent it

[5] Quoted in E. R. Pike, *op. cit.*   [6] *op. cit.*

25 Women were employed underground in British mines until the Mines Act of 1842 was passed

were not swift. For example, a decree in Prussia of 1839 forbade the employment of children under nine in mines and factories, and limited the labour of younger people under sixteen to ten hours. Not until 1882 was the employment of children under the age of twelve forbidden.

The development of the coal industry to the mid-nineteenth century had been centred on Britain; as yet the coal-producing countries of Europe could not challenge her supremacy. British and Belgian miners had led the way in overcoming the problems hindering the rapid expansion of mining. Coal production had increased enormously especially in Britain, although the miner had not shared the benefits, and this expansion stimulated the development of canals and railways. The second half of the century was to witness an even greater expansion of production and the emergence of America and Germany as two formidable rivals to Britain's position as the world's leading coal producer.

4 Developments in the mining of base metals from the eighteenth century to the mid-nineteenth century

UP TO THE MIDDLE of the eighteenth century the mining of metals such as iron, copper, tin, and lead was confined largely to Europe and was often carried on in areas where they had been mined since medieval times, such as the copper mines of Falun in Sweden and the tin mines of Cornwall. Demands for such metals, which had been limited to the needs of agricultural communities and of war, rose as new uses were found for them, particularly for iron in the making of machinery, bridges, buildings (especially to strengthen factories), and railways. Output of metal mines increased steadily and was linked with solution of problems which had previously hindered production.

Eighteenth-century metal mines were faced with many of the same problems as coal mines and the most serious was that of removing water from the workings which, if unsuccessful, compelled some mines to close. Celia Fiennes, who made her *Great Journey* into Cornwall in the autumn of 1698, noted the enormous efforts made to keep tin mines free from water.

> There were at least 20 mines all in sight, which employs a great many people at work, almost night and day, but constantly all and every day, including the Lord's day, which they are forced to, to prevent the mines being overflowed with water.

Miners constructed adits to clear the mines, some of which were very complex; for example the Nent Force Level near Alston in Cumberland, which was constructed by Smeaton from 1776, nearly 5 miles long, and able to drain a wide area of ground. It was not always possible to construct a successful adit, as W. Pryce noted in 1778 in his book *Mineralogia Cornubiensis*:

26   Falun copper mine, Sweden, in the eighteenth
century. The largest copper mine in northern
Europe, it was worked from the thirteenth century
onwards, and this picture shows the scale of the
enterprise at the surface

27   A Cornish copper mine drained by a steam
pumping engine, the beam of which can just be
seen protruding from the second storey. The
capstan on its left was probably used for raising
pump rods when necessary

53

28 The vast disused Parys copper mine, Anglesey, closed in the nineteenth century. It had been worked since Roman times but it could not compete with cheap American copper

> . . . with all the skill and adroitness of our miners they cannot go to any considerable distance below the adits before they must recourse to some contrivance for clearing the water from their workings.

Machines to drain mines were not unlike those described by Agricola, although hydraulic engines were introduced in Germany in about 1748 and in England in the late eighteenth century. The development of the steam engine, especially of Boulton and Watt engines, was of enormous value to metal mining. Their superiority over Newcomen engines was demonstrated in 1781 when Consolidated Mines of Cornwall replaced seven Newcomen engines with five Boulton and Watt engines and saved £10,830 per year in coal bills alone. However some mines continued to use the hydraulic engine, especially in Cornwall and northern England in conditions where steam engines would have been too expensive to maintain.

The developments in winding machinery for men and materials which took place in coal mining occurred in metal mining too. The most interesting innovation was the man engine which was common in Cornish mines.

54

29   German iron ore miners in the Harz Mountains working in a series
of ascending levels known as overhand stoping

A surface pumping engine transmitted motion by rods down a shaft to a
stationary pump near the bottom. These rods had a stroke of 6 to 12 feet,
with a pause at the end of the stroke while the valve gear reversed at the
pumping engine. At the side of the shaft, opposite the point of the end of
the stroke of each rod, and spaced the distance of the stroke apart, were
placed strong platforms to hold a man. At corresponding points on the
pump rods smaller platforms were fixed. If a miner wanted to go down the
mine he stepped on the platform attached to the rods. As the rod descended
the miner was carried down and when it stopped at the end of the stroke
the miner stepped on to the convenient platform fixed to the shaft. The
rods rose again, stopped, and the miner stepped on to the next platform on
the rods and was carried down a further stage when the rods descended.
Man engines like this were installed particularly in Cornish mines in the
early nineteenth century but they were to be found all over the world
as Cornishmen emigrated when the mining industry collapsed in Cornwall
at the end of the century. There is an old mining maxim, 'Cornishmen do
not like to hang to a rope', no doubt referring to their preference for the
man engine.

55

The most outstanding difference between coal and metal mining was in the method of digging out what was being mined. This is accounted for by the fact that deposits of coal frequently lie horizontally beneath the surface while mineral deposits often lie vertically, although iron ore often lay horizontally and was mined in 'bell pits' even in the eighteenth century. In metal mines the most important feature was to follow the vein underground. The first operation consisted of driving a main shaft, usually rectangular in the eighteenth century, into the ore body. Then a series of horizontal levels were driven from the main shaft and a series of smaller connecting shafts from level to level. These smaller shafts were to provide routes for the subsequent removal of the mined ore. The method of mining was overhand or underhand stoping, that is working in steps. To open a new underhand stope a pillar of solid ground was left below an upper level carrying the road for haulage and the travelling road for the workings above. A slice of the lode was taken right and left from one of the connecting shafts. The ore was broken, sometimes by gunpowder, then shovelled into a connecting shaft where it fell to the level below and was transported to the shaft. When the first levels had travelled a few yards, two more were started below them, so that in time there was a series of descending levels, like a series of huge steps from the connecting shaft. In overhand stoping the miners began to work at the bottom of the connecting shaft, immediately above the main level and drove into the ore body horizontally, rarely vertically, and thus there was a series of ascending levels.

Eighteenth-century miners used tools similar to those described by Agricola; the pick, the crowbar, wedge, and shovel. If the rock was particularly hard fire setting was used in some areas. Many mines used gunpowder, which was the case in Britain by 1689 as the parish register of Breage indicates in this entry:

> Thomas Epsly of Chilcumpton parish, Sumersitsheere. He was a man who brought that rare invention of shooting the rocks which came heare in June, 1689, and he died at the bal [i.e. mine] and was buried at the breag on the 16 day of December 1689.

By the eighteenth century the miners had devised a primitive safety fuse which cut down on the number of accidents. Holes 3 to 4 feet in depth and about 3 inches in diameter were bored by hand and packed with gunpowder. Into the gunpowder a small circular copper bar was inserted and the remainder of the hole filled with soft clay, and then the copper bar was

30 Nineteenth-century washing pans of wood at a Welsh lead mine. The lighter rock was washed away, the heavy galena remaining behind

31 The disused engine house of Levant tin mine, Cornwall, stands as a grim reminder of the collapse of the Cornish tin industry in the nineteenth century

withdrawn and replaced by a straw. The straw filled with gunpowder was lit by a piece of touch paper or a slow match, the miners withdrew to a safe distance and the subsequent explosion loosened the rock which could be shovelled down the connecting shafts.

Serious efforts were made to develop a drill that would make the task of boring the rock easier. The boring bars used to about the middle of the nineteenth century had been of iron bar with a 'steeled' end which did not stand up for long to hard rock. About 1850 cast-steel rods for boring rock, which were more durable and speeded up the rate of drilling, were introduced into Derbyshire lead mines. In 1813 Richard Trevithick developed a hand operated rock drilling machine, but a power drill was not developed until after the middle of the century. When the Mont Cenis tunnel was constructed an engineer called Bartlett made a steam-powered boring machine which was used with some success. The development by an engineer called Sommeille of a drill driven by compressed air, which was also used in constructing the Mont Cenis tunnel, turned out to be still more successful. The principle on which these early machines were based was the boring of a hole either by continuous motion of a rotating drill or by intermittent blows delivered by a pointed tool striking a rock. Low's boring machine was the most successful of the early inventions since it attempted to combine both of these operations. It consisted of a cylinder into which a tool similar to the hand-held chisel was inserted. This cylinder was housed in another cylinder in which it was made to rotate slightly and continuously between each blow of the drill, the latter striking between 300 to 500 blows per minute. The machine was carried on a trolley and driven by compressed air. In 1863 Alfred Nobel invented dynamite and its use in cartridge form gave a much greater rock-breaking power than was the case with gunpowder.

From the 1860s onwards metal mines in Europe, but particularly British mines, began to suffer from intense competition from mines which had been opened up in America and South America. Copper, tin, and lead mines were badly affected. The best lodes were nearing exhaustion in many parts of Europe whilst particularly rich reserves were found in America, especially copper on the Keweenaw peninsula on Lake Superior. In fact from this time onwards metal mining tended to be dominated by companies operating outside Europe. The talents of the European miners were not lost, many of them, particularly Cornishmen, turned up in large numbers in mining enterprises that developed in many parts of the world. It would be inaccurate to suggest that this decline affected all metal mining; in fact the mining of iron ore continued to be an important industry in many parts of Europe, but the major technical innovations, especially in open cast mining, occurred outside Europe.

5 Gold mining in the nineteenth century – the age of the Lone Prospector

UNTIL THE MIDDLE of the nineteenth century gold was in very short supply; a series of spectacular gold rushes from 1848 onwards ended this situation. The chief gold producers until 1848 were countries in Central and South America and Russia. European production, except in Russia, was very low, gold frequently being recovered as a by-product from base metal ores such as copper. Nineteenth-century gold rushes have enormous human interest because the individual prospector could make his fortune. The same cannot be said for the Russian goldfields that yielded three-fifths of the world's gold until 1848.

The rapid development of Russian gold mining began in 1744 with the discovery of a quartz outcrop near Ekaterinburg. This mine proved profitable but it was to be eclipsed by the numerous alluvial goldfields discovered near this town in the early nineteenth century. Gold was discovered in Siberia in 1838 on the river Ulderey in the Yenisey basin and led to a gold rush in the 1840s. By 1851 about 20,000 miners were working in 106 gold mines of the district and up to the 1870s this was the biggest goldfield in Russia, producing 20 per cent of the country's yield. The most significant mining area after this was the Lena basin goldfield where mining began in 1846. The greatest development took place on the river Vitim, the settlement of Bodaybo becoming the main focus. The miners were all Russians and they formed a distinct group of settlers. They were mainly unskilled and recovered gold by primitive methods such as panning the gold. Some miners were convicts, but many more were exiles who could find no better employment. These northern mining areas, unlike those in southern Siberia, were not state-run with forced labour but were private undertakings which obtained labour where they could. The harsh climate of Siberia frequently took its toll on the workers, as one traveller pointed out: 'One must have the iron constitution of Siberia to bear such

59

fatigue and privations; but even of them many succumb.'[1] Russian gold production, however, was totally eclipsed by spectacular gold discoveries in America, beginning with that in California in 1848.

The origins of the Californian gold rush began on a January morning in 1848 when James W. Marshall found what he thought were specks of gold in the tailrace of a sawmill under construction for John A. Sutter on the American river, some 40 miles above its junction with the Sacramento. His own account captures best his excitement on that clear, cold January day:

> . . . I shall never forget that morning – as I was taking my usual walk along the race, after shutting off the water my eye was caught by a glimpse of something shining in the bottom of the ditch. There was about a foot of water running there. I reached my hand down and picked it up; it made my heart thump, for I felt certain it was gold. The piece was about half the size and of the shape of a pea. Then I saw another piece in the water . . . I thought it was gold, and yet it did not seem to be of the right color . . . Suddenly the idea flashed across my mind that it might be iron pyrites. I trembled to think of it! This question could soon be determined. Putting one of the pieces on hard river stone, I took another and commenced hammering it. It was soft and it didn't break; it therefore must be gold . . .[2]

Sutter was interested in completing his sawmill and tried to keep the men at work. This he succeeded in doing until March when several workers deserted him and moved off to mine 25 miles away at a spot later known as Mormon Diggings. San Francisco was a struggling port of some 2,000 inhabitants and the news of these discoveries was brought by one Sam Brannan who is reported to have rushed through the streets of the town with a bottle of dust in his hand shouting, 'Gold! Gold! GOLD from the American river.' Within weeks the territory of California was swept by a gold fever; the impact is vividly captured in the *San Franciscan Californian* of 29 May 1848:

> The whole country from San Francisco to Los Angeles and from the seashore to the base of the Sierra Nevadas resounds with the sordid cry of GOLD, GOLD!, the field is left half planted, the house half built, and everything neglected but the manufacture of shovels and pickaxes!

[1] T. Green, *The World of Gold*

[2] Quoted in H. S. Commager and A. Nevins, *The Heritage of America*

Monday 24th this day some kind of mettle was was found in the tail race that that looks like goald first discovered by James Martial, the Boss of the mill. Sunday 30 clean & has been all the last week our metal has been tride and proves to be goald it is thought to be rich we have pict up more than a hundred dollars worth last week

February. 1848
Suny 6th the wether has been clean

32   An entry in the diary of a Mormon labourer at
Sutter's Mill describing the discovery of gold

Sordid it may have been in the opinion of this newspaper editor, but the earliest miners were not disappointed in the yields. Two men took $17,000 in dust and nuggets from one canyon; in eight days five men at Mormon Diggings made a profit of $1,800. On the whole, miners in 1848 averaged about one ounce of dust daily, worth about $20, good earnings when labourers in eastern America were paid $1 a day.

Once news of these profits reached eastern America the rush to California began, especially as excitement was raised by the wildest stories of fortunes that were made. There were three routes to California; by ship round the

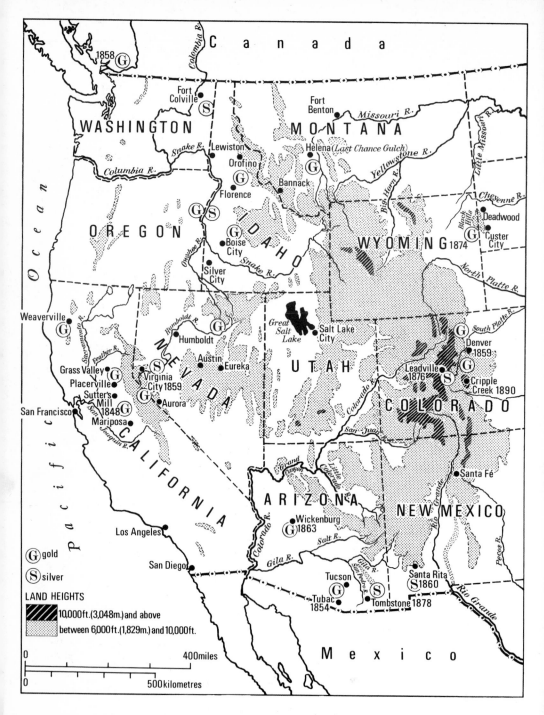

C a n a d a

1858 Ⓖ

Fort
Colville Ⓢ

WASHINGTON

Columbia R.

Lewiston Ⓖ
Orofino Ⓢ
Florence
Bannack

Snake R.

Columbia R.

MONTANA

Fort
Benton

Missouri R.

Helena *(Last Chance Gulch)* Ⓖ

Yellowstone R.

OREGON

Ⓖ Ⓢ

I D A H O

Ⓖ
Boise
City

Silver
City

Snake R.

Owyhee R.

WYOMING 1874

Big Horn R.

Cheyenne R.

Black Hills

Deadwood
Custer
City

Little Missouri R.

O c e a n

Weaverville Ⓖ

Sacramento R.

Feather R.

Grass Valley Ⓖ
Placerville
Sutter's
Mill
1848 Ⓖ
Mariposa

San Francisco

San Joaquin R.

Humboldt Ⓖ

Humboldt R.

Austin
Ⓢ Eureka

N E V A D A

Virginia Ⓢ
City 1859
Ⓖ Aurora

Great
Salt
Lake

Salt Lake
City

U T A H

Colorado R.

San Juan R.

Denver
Ⓖ 1859

Leadville Ⓢ
1876 Ⓖ

Ⓖ *South Platte R.*

North Platte R.

Cripple
Creek 1890

C O L O R A D O

Santa Fé

Rio Grande

Pecos R.

C A L I F O R N I A

Los Angeles

San Diego

Colorado R.

Grand Canyon

Little Colorado R.

Gila R.

A R I Z O N A

Ⓖ Wickenburg
1863

Salt R.

Tucson

Tubac Ⓖ
1854

San Pedro R.

Gila R.

Ⓢ Tombstone 1878

N E W M E X I C O

Santa Rita
Ⓢ 1860

Rio Grande

M e x i c o

Ⓖ gold
Ⓢ silver

LAND HEIGHTS

▨ 10,000ft.(3,048m.) and above

░ between 6,000ft.(1,829m.) and 10,000ft.

0 ⊢──────────────┤ 400 miles

0 ⊢──────────────┤ 500 kilometres

MAP NO. 3   MAJOR GOLD AND SILVER FINDS IN AMERICA, 1848–90

Horn, by ship to Panama, then across the isthmus on a donkey and by ship on to San Francisco, or, finally, the long route using the overland trails. Casualties along all routes were high; survivors endured incredible physical hardships to reach California. Fever killed many at Panama, others died on overcrowded boats negotiating the Horn, whilst as many as 5,000 were killed by cholera, Indians, and accident crossing the Great Plains. The dangers of the long overland crossing are emphasized by Meek, a guide to a party of 'forty-niners':

> No, it's not high mountains ner great rivers ner hostile Injuns that'll give us most grief. It's the long grind o' doin' every day's work regler an' not let-up fer nobody fer nothin'. Figger it fur yourself; 2,100 miles – four months to do it in between April rains and September snows – 123 days. How much a day and every cussed day?
> Yaas, and every day rain, hail, cholera, breakdowns, lame mules, sick cows, washouts, prairie fires, flooded coulees, lost horses, dust storms, alkali water. Seventeen miles a day – or you land in the snow and eat each other like the Donner party done in '46.[3]

Most of these early miners had little knowledge of mining, but none was needed in the early days. Gold was found in dry beds of former streams, gravel was shovelled out until they reached bedrock ,where they frequently found pockets of gold that could be lifted out by the handful. This was then separated from dirt by tossing it in a blanket while the wind blew away the dirt, a very wasteful method. 'Dry washing' did not persist for long. Many miners used a technique known to the earliest miners in history, washing out the gold from gravel in a stream. Gold could be separated from gravel and sand by using a washing pan, a shallow metal vessel into which 'pay dirt' was thrown; this was lowered into a stream and gently rotated, leaving the gold in the bottom of the pan. Isaac Humphrey, who introduced the pan, also built the first 'cradle' in California in 1848. This simple device was described by J. Fremont, in his *Geographical Memoir* dated 1849:

> . . . the greater part had a rude machine, known as the cradle.
> This is on rockers, six or eight feet long, open at the foot, and at its head has a coarse grate or sieve; the bottom is rounded, with small cleats nailed across. Four men are required to work this machine; one digs the ground in the bank close by the stream; another carries it to the cradle and empties it on the grate; a third gives a violent rocking motion to the machine; whilst a fourth dashes on water from the stream itself. The sieve keeps the coarse stones from entering

[3] Quoted in Alan Lomax, *The Folk Songs of North America*

33　A group of Californian prospectors using a cradle similar to that described by J. Fremont in his *Geographical Memoir*

the cradle, the current of water washes off the earthy matter, and the gravel is gradually carried out at the foot of the machine, leaving the gold mixed with a heavy fine black sand above the first cleats.

The sand and gold mixed together are then drawn off through auger holes into the pan below, are dried in the sun, and afterwards separated by blowing off the sand.

Not long after a more sophisticated device was introduced – the 'long tom'. It consisted of two long boxes placed one on top of another. Dirt was thrown into the top box, water running through it carried away gravel and soil while the heavier gold and sand fell through holes in a perforated iron floor. More water passed through the bottom box washing out the sand and leaving the gold on the ridged bottom. The earliest 'long toms' were 12 feet long; by 1850 some were 100 feet long and required a great many workers plus a very plentiful supply of water.

Mining camps sprang up wherever a promising new area was found. They attracted men from all over the world. Americans were in the majority but other nationalities were plentiful; there were 25,000 Frenchmen and nearly 20,000 Chinese by 1853. Life in the mining camps was tough; the murder rate in the territory was terrifying. Hard physical labour in all weathers was followed by a meal of pork, beans, and bread in some ramshackle hut. Diarrhoea, dysentery, and scurvy were all common. Drinking and gambling were the usual leisure occupations; the saloons, so much a part of the legend of the West, provided the venue. Riches were not the reward of all miners. In the early days they sang cheerfully,

> Oh Susannah, don't you cry for me
> I'm gone to California with my wash bowl on my knee

but many were sadly disillusioned,

> It's four long years since I reached this land
> In search of gold among rocks and sand
> And yet I'm poor when the truth is told
> I'm a lousy miner
> I'm a lousy miner in search of shining gold.

34   Miners panning for gold in California, 1853

Yet unsuccessful miners had a habit of turning up in other places where strikes were made; there were plenty in the 60s in the North American continent.

The Californian gold rush was the first of a spectacular series of gold finds in North America in the nineteenth century. The process was to be repeated time and again: in Colorado, Nevada, Arizona, Idaho, Montana, British Columbia, Dakota. The Nevada finds, especially the area known as the Comstock Lode, were perhaps the most important after the Californian. Henry J. Comstock, after whom the area was named, was a claim jumper who died penniless, but the men who reaped much of the reward were John Mackay, James Fair, James Flood, and William O'Brien, who in 1873 discovered a particularly rich ore vein in the Consolidated Virginia Mine. The profits from this mine were enormous; it was valued at 75\frac{1}{2}$ million in January 1875 and one share worth $1 in 1870 sold for $700 in January 1875. The Nevada gold rush had all the features evident in preceding and subsequent rushes; the high hopes of the early arrivals, bitter disappointment for most, mining conducted under harsh physical conditions – plus one further very significant factor, the fairly rapid exhaustion of alluvial gold and the development of quartz mining. The latter meant removal of the rock in which the gold was located, crushing the ore body and then removing the gold. It also meant that mining moved underground and consequently high capital cost. It is significant too that as much silver as gold was obtained from the Comstock and this in itself posed problems, since silver is usually found in association with base metals (lead, zinc, or copper) and can be separated from them only by the costly and exacting process of smelting. So, into the goldfields moved the capitalists, the men with money who could finance the deep mines and pay for the expensive surface machinery to crush the ore and remove the gold. This development spelt doom for the small group of prospectors since few had the capital for such ventures. After the initial rush the majority of miners became wage earners who shared none of the fortunes that were made. The Consolidated Virginia Mine in the Comstock illustrates this point. The richest ore vein was located at a depth of 1,200 feet; its produce was enormous but the 1,000 or so miners working in it in 1877 at the height of its production were wage earners, the profits went to the shareholders. The most famous name associated with Nevada was George Hearst of San Francisco who invested heavily in the Comstock. He did likewise in the Dakota strike of 1877. Once the alluvial gold had been taken he and his fellow Californian capitalists

35 San Francisco in 1851, when it was beginning to grow rapidly owing to the gold rush – yet it still looks very much like a frontier town

moved in on the bonanza, purchased the richest mines, and reaped considerable financial benefits from their investments.

The early mining towns that mushroomed round these gold strikes all exhibited similar features. Nathaniel Langford's description of one such town, Virginia City, Montana, vividly recalls life in a mining community and can be taken as typical of the mining towns of the West, as seen through the eyes of a contemporary observer:

This human hive, numbering at least ten thousand people, was the product of ninety days. Gold was abundant, and every possible device was employed by the gamblers, the traders, the vile men and women that had come with the miners to the locality, to obtain it. Nearly every third cabin in the towns was a saloon where vile whiskey was peddled out for fifty cents a drink in gold dust. Many of these places were filled with gambling tables and gamblers, and the miner who was bold enough to enter one of them with his day's earnings in his pocket seldom left until thoroughly fleeced. Hurdy-gurdy dance houses were numerous, and there were plenty of camp beauties to patronize them. Not a day or night passed which did not yield its full fruition of fights, quarrels, wounds or murders. The crack of the revolver was often heard above the merry notes of the violin. Street fights were frequent . . .

Sunday was always a gala day . . . Thousands of people crowded the thoroughfares, ready to rush in any direction of promised excitement. Horse racing was among the most favored amusements. Prize rings were formed, and

36 Alluvial gold was quickly exhausted in Australia and mines like this one in Victoria, needing a lot of capital to develop them, came into operation (1859)

brawny men engaged in fisticuffs until their sight was lost and their bodies pommelled to a jelly, while hundreds of onlookers cheered the victor . . .[4]

The opportunities for the individual prospector were not confined to North America. In Australia, the rush was initiated by Edward Hargreaves, an unsuccessful digger in California who returned to Australia in 1850 determined to find gold. Hargreaves's determination bore fruit for he discovered gold at Bathurst in New South Wales within a few weeks of his return. The rush that followed was repeated in 1851 in Victoria when gold was discovered at Ballarat, a mere 60 miles from Melbourne, and the impact on Victoria was as great as Marshall's discovery had been on California, as Governor Charles le Trobe's report to the British government indicated:

Cottages are deserted, houses to let, business is at a standstill, and even schools are closed. In some suburbs not a man is left . . .

[4] Quoted in H. S. Commager and A. Nevins, *op. cit.*

Le Trobe's report in the following year indicated the same features of mining life as noted in America:

> . . . the number of thoroughly dissolute characters has increased . . . and this great fact, taken into account with the great increase . . . of the illicit sale of spirits . . . is sufficient to account for any amount of disorder . . . Violent quarrels, thefts among the huts, tents and workings have been common.

Nevertheless, le Trobe's despatch of 2 March 1852 to the Secretary of State indicated that the authorities had a considerable measure of control over the diggings and were aware of the enormous problems confronting them in keeping law and order.

Shortage of water restricted the use of the long tom but skilled miners supplemented the rocker by the puddling tub, made out of half a barrel, to break down and dissolve the stiff Victorian clay. The tub was half filled with pay dirt, water was baled in from the creek or river, and the contents worked with a spade until the clay dissolved in the water. As the water became heavy with earth it was poured out and a fresh supply added. Eventually nothing would remain but clean gravel, sand, and gold.

As the alluvial gold was exhausted it became necessary to sink fairly deep shafts, especially in the Ballarat area, in order to reach the gold-bearing ore. Inevitably the individual miner could not compete against the mine companies and he gradually disappeared in spite of the introduction of a modification of the Cornish practice of 'tributing'. Ground would be let for a term of five years or less, to a group of miners for a percentage varying from 10 to 30 on the gross quantity of gold extracted from the mine. The mine proprietors were sleeping partners in spite of the strict rules they imposed. Gradually, however, as in America, the individual miner had no place in gold mining once it became necessary to sink deep shafts, install machinery, and hence expend considerable sums of money. Nevertheless, along the eastern coastal strip of Queensland there was plenty of opportunity for the individual prospector in the late 1860s and the 1870s, because there was a series of discoveries ranging from Gympie, just north of Brisbane, to as far north as the Palmer river in tropical Queensland. In the 90s the yield of the Queensland fields was at its height but by this date many prospectors had been attracted to Western Australia, particularly to Coolgardie and Kalgoorlie.

Western Australia was the location of Australia's greatest gold strike

69

37　This tented camp in the Coolgardie area in
1895 was typical of many that had sprung up once
gold had been discovered in Australia

when in 1892 Arthur Bayley and William Ford, two tough, experienced
miners and prospectors, discovered gold in the arid, semi-desert area
around Coolgardie. Shortly afterwards Paddy Hannan, an Irishman with
long experience in the goldfields, discovered gold some miles from Coolgardie at a place later to be known as Kalgoorlie. The rush did not last
long, since alluvial gold in large quantities was not found, the gold-bearing
lodes were not easy to locate, and expensive machinery was needed.
Hannan made no fortune out of his discovery but this was not unusual
among these lone prospectors. He died in poverty, as had James Marshall.

The Yukon and Alaskan gold rush of 1897–98, the last gold rush of the
nineteenth century, had all the glamour, heartbreak, and excitement of the
first. Two prospectors, Robert Henderson and George Washington
Carmack, were salmon fishing in a tributary (to be known later as the
Klondike) of the Yukon river in northern Canada one August afternoon in
1896, when the gleam of gold caught their eye. This was not the first gold
to be found in the far north; the best-known source was near a town called
Circle City in the United States territory of Alaska. News of this find did
not reach the rest of the North American continent until the spring of 1897,

38　An endless stream of would-be prospectors
climbing the Chilkoot pass to the Yukon (1897)

and then a rush began that recaptured the fever of California and '49,
'Klondikitis' as the newspapers called it. Dawson City rose from nothing
as the gold mining centre; within a year it had a population of 25,000. By
February of 1898 forty-one ships were on a regular run from San Francisco
to Skagway, the nearest port to the goldfield. From this port the prospectors had to cross either the Chilkoot Pass or the White Horse Pass. The
Chilkoot had the advantage of being more direct and was more popular
during the winter of 1897–98, when 22,000 people were checked through the
boundary station at its summit by the North West Mounted Police (the
boundary between Canada and the U.S.A. was at the top of the pass). It
was not simply a matter of man taking his place in the 4 mile long climb
from Sheep's Camp to the summit of the Chilkoot. Each was required to
bring with him a year's supply of food, about a ton of goods in all, and a
handful of the Canadian North West Mounted Police refused entry to
prospectors who did not fulfil this requirement. Once at the top, a 500 mile
river journey was necessary to reach Dawson City, a journey that could not
be undertaken until the winter snows had melted. Few could reach their
destination until June 1898.

39 The primitive nature of living and working
conditions is shown in these photographs of
Klondike in the first year of the rush (1898)

The survivors of this terrible journey to Dawson City could expect to find it worthwhile. Some underground mining was undertaken, but the first miners concentrated on extracting gold near the surface. The summers were brief, the winters long and bitterly cold. The season of low water during which gold in river beds could be worked was short. The gold-bearing gravel in the gulches was in ground frozen hard to a depth of 18 inches under a thick coating of moss. Miners burnt off the moss but only shallow rich deposits could be worked profitably throughout the short summer; when winter came mining ceased, until the practice of drifting or fire setting was adopted. Throughout the long winter months the ground was thawed nightly by fires and the gold-bearing dirt was put to one side. In summer the resultant dumps were put through sluice boxes such as the cradle and long tom. Drifting or fire setting proceeded at the rate of about a foot a day; work could go on throughout most of the year. This technique represents a triumph of the miner's adaptation to the conditions. The rush yielded about $2\frac{1}{2}$ million of gold in the first three years; by 1900 the most accessible gold had gone. Once again, heavy capital investment was needed to mine gold. An example of such expenditure was the work of the Yukon Gold Company organized by A. N. C. Treadgold. Visiting the area in 1906; he saw the possibility of rewashing gravel with powerful jets of water if a large and steady water-supply could be obtained. He acquired many worked-out claims, brought water 70 miles from the Twelve Mile river, and he and his associates reaped rich rewards.

Mining towns were as colourful in Canada as in America, and Dawson City was no exception. Miners lost their hard-earned gold in its dance halls, the most famous being the Palace Grand built by 'Arizona Charlie' Meadows, formerly with 'Buffalo Bill' Cody's Wild West Show. Here the miner might meet women like famous 'Snake-Hips Lulu', who would have been anything but easy to our eyes, but to a miner working in the harsh climate of the Yukon must have had some appeal. Not all the women belonged to the dance halls; there were hard-working wives as well, some of whom helped to pack supplies across the passes. In the early days there were even four nurses of the Victorian Order, under the careful protection of the North West Mounted Police.

No history of these gold rushes is complete without looking at their impact on the countries concerned, and their effect on the world at large. The rushes occurred in areas that were either undeveloped and under-populated, or had no population at all. California before 1848 fell into the former category; her population explosion was phenomenal. Before the end of 1852, the peak year of the Californian gold boom, the state had a quarter of a million inhabitants (over fifteen times as many as at the beginning of 1848); by the census of 1860 population had risen to 380,000. Between the expanding state of California and the developed eastern states lay the vast empty area of the Great Plains, long known as the 'Great American Desert'. One of the fascinating features of American history in the period 1850–90 is the conquest of the Great Plains. Many factors brought about this conquest, but the gold rushes, ephemeral though some of them were, played an important part. Precious metals were the magnet that drew people to Colorado, Nevada, Arizona, Idaho, Montana, and Wyoming. As the gold was exhausted the mining population receded and its place was taken by ranchers and farmers who established, with the aid of the railroads and the government, the permanent foundation of the territory. Some miners became farmers as *The Boise Statesmen* of Idaho dated 1 July 1870 indicated:

> As the placer mines decline persons forsake them for the more permanent pursuits of farming and stock-breeding . . . the grain, hay and vegetable crops of Boise and other agricultural districts is now better than ever.

The miners familiarized the American people with the country between the Missouri and the Pacific. They focused attention on the Indian problem and drew attention to the need for railroads. Gold mining contributed enormously to the development of towns, particularly San Francisco. Every advance of the Western mining frontier lent prestige to its mint and to its stock exchange, and expanded its port facilities, while advancing the fortunes of the rich.

The early gold rushes in Victoria and New South Wales brought great changes to Australia. In 1850 there were 400,000 people in the Australian colonies; by 1861 there were nearly three times as many. Wheat growing and sheep farming suffered from the gold fever, when many men left farms to go to the diggings. But after the first check the increased demand for food caused more land to be cultivated, and the area under crop more than doubled – 480,000 acres in 1850, over one million in 1860. Demand for meat caused the expansion of the cattle and sheep industry. Once the initial rush was over many diggers wanted to settle in the towns and Melbourne and Sydney both expanded considerably; the former becoming

40 Placer mining in the Klondike. Gold-bearing gravels were washed out using a powerful jet of water. A similar technique, sometimes called hushing, was used by the Romans in Spain

the leading financial centre of Victoria. They did less than the American rushes to 'open up the country'. The earlier rushes were in country already opened up by the squatters, and hardly anywhere did the gold, as in the Rockies, prove to be an indication of even more important deposits of base metals. Perhaps the most interesting result of the Australian gold rushes was to put an end to the transportation of convicts to Victoria and New South Wales. The Chief Secretary of New South Wales was absolutely correct when he remarked on hearing of Hargreaves's discovery, 'If this is a gold country, it will stop the Home Government from sending us any more convicts.' Transportation was abolished, except to Western Australia. This state, a part of it previously without population, saw the growth of two thriving towns, Kalgoorlie and Coolgardie.

The Yukon and Alaskan gold rush opened up a virgin territory. The discoverers found gold in an almost uninhabited wilderness, an outpost of modern civilization, and an outpost it has remained. Recent oil discoveries in Alaska may well change this state of affairs. It is necessary to look elsewhere for the chief effects of Klondike and Alaska. The rush was one factor in the great development of western Canada, particularly in the growth of British Columbia. The effect on the scattered towns of south east Alaska was significant too. America neglected the territory until the gold rush. It was not by mere coincidence that Alaska received a criminal code in 1899, a civil code and a municipal government law in 1900. The impact of the gold rushes is not to be left here – all added to the folklore of the countries concerned, enriched their idiom, and contributed to their literature. The Yukon and Alaskan rush gave Jack London the material for at least a dozen novels; it inspired Robert Service to write *The Shooting of Dan McGrew* and other poems; it provided the background for Charlie Chaplin's film *The Gold Rush*. More recently the Californian gold rush provided the inspiration for a highly successful American musical, *Paint Your Wagon*.

The rushes led to an enormous increase in world output. In the whole of the first century after Columbus discovered America the world output of gold was about 750 tons; in the second half of the nineteenth century it was 10,000 tons. Such a rise had an impact on the major European powers. Gold was used more and more in business transactions. Britain in 1850 was the only country on the gold standard, a system by which the value of money is related to a fixed weight of gold; the 1870s saw most of the major European countries (except Russia) adopt a gold standard. By 1900 almost every country in the world had changed to a gold standard, except China.

This change to the gold standard was made possible by the gold discoveries in the second half of the nineteenth century, in which the lone prospector was the key figure in setting in motion the process. Twentieth-century gold mining has no place for him; gold mining is dominated by the corporation relying on scientific prospecting and needing enormous capital investment to mine the gold once it is discovered. The 'forty-niner' with his plaintive song is a distant figure in the gold mining industry of the twentieth century:

> O land of gold, you did deceive me,
> And I intend you my bones to leave,
> So farewell home, now my friends grow cold,
> I'm a lousy miner,
> I'm a lousy miner, in search of gold.

6 Coal mining since the mid-nineteenth century – coal the 'King' dethroned?

THERE HAVE BEEN three very important developments in coal mining since the mid-nineteenth century; the emergence of America and the U.S.S.R. as the leading coal-producing countries, the mechanization of the process of digging coal, and the challenge that new fuels have offered to the coal industry. Up to the mid-nineteenth century Britain's position as the leading coal-producing country was unchallenged, but by the beginning of the twentieth century Britain had been ousted from first place.

Coal production in 1912 in Tons[1]

America	477,202,000
Britain	260,416,000
Germany	172,065,000
France	39,745,000
Belgium	22,603,000

Coal mining as an industry can hardly be said to have existed in America until after 1820. In 1821, the earliest year for which a record is available, the total output was only 1,322 tons. Production increased slowly, but more rapidly after 1850, and from 1870 to 1910 each decade virtually doubled, a peak being reached during the First World War. During this century America has been overtaken in coal production by the U.S.S.R., a striking testimony to the high degree of industrial development of that country since the revolution of 1917 and the domestic upheavals that followed. During the period 1957 to 1961 the U.S.S.R. produced an average of 460 million tons of coal a year. America produced an average of 445 million, and Britain of 225 million.

The most obvious feature of the change in mining methods has been the replacement of the pick and shovel by the coal cutting machine, and more recently the introduction of automated machinery on the coal face. Equally

---

[1] H. Stanley Jeavons, *The British Coal Trade*

interesting is the development of open pit coal mining, a technique first developed in the mining of lignite or brown coal, but used increasingly in areas of bituminous coal, especially in America. Whilst total world production of coal continues to rise the industry faces serious competition in some countries from the development of other forms of energy such as oil and natural gas. Inevitably the question arises whether coal will lose its first place as a source of energy.

In the technique of digging coal the most important development has been the introduction of machinery to undercut the coal, the most difficult task of the hewer. In this field American mining took the lead, partly because the seams were thick and easily accessible, and partly because some coal owners, particularly the British in the early twentieth century, showed little enthusiasm for such innovations. Yet the early pioneering work in this field was done by British engineers. Cutting coal by machine rather than by hand was first recorded in 1761 when a device called 'Willie Brown's Iron Man' was used; it wielded a pick like a man, striking the coal face with harder and more frequent blows. Early in the nineteenth century a machine was constructed on a similar principle to the 'Iron Man' which used a horse to supply the necessary power. Other experiments involved the use of circular steel discs to cut the coal like a rotating knife blade. All these machines had a common fault in that they lacked a suitable form of power since they relied on manpower or horsepower to work them. Without effective power they were bound to have a very limited use.

The problem of power was solved by the introduction of compressed air, a discovery credited to a British aristocrat called Lord Cochrane. By 1850 compressed air power was beginning to replace steam power for many underground purposes in British and European mines. As a medium for the transmission of power to machinery in underground workings, compressed air was very suitable. It could be conveyed easily to the most distant workings, there was no heat in the pipes and no reason to fear explosions, and the air released when machinery was in use helped mine ventilation considerably.

A variety of coal cutting machines using compressed air were introduced, the first successful one in 1861. The most popular type developed were the disc machines. Undercutting the coal was done by means of a disc, 3 feet to 6 feet in diameter, on the edge of which were fixed a number of picks or cutting tools. The disc worked the same way as a circular saw but was placed horizontally instead of vertically. The first successful one

78

41   The first disc coal cutter, manufactured in
Wigan, 1868

was used in 1868 on the Lancashire coalfield. American engineers played a
very important part in the development of coal cutting machines; in 1894
they developed a successful chain machine. The method was to undercut
the coal with a continuous chain that revolved round a long arm attached
to the machine. A bar machine introduced in the 1860s consisted of a
circular tapered steel bar with a number of teeth or cutters fixed to it. It was
especially valued in seams which were soft and friable since only a small
area of the seam was undercut and left unsupported for a very short time.
A disc machine was often hindered in seams like these since it undercut a
lot of the seam, the soft coal fell on to it, and work had to be stopped until
it could be cleared. These machines were at first powered by compressed
air but early in the twentieth century machines worked by electricity were
introduced. Electric coal cutters made slow headway against suspicion
(sometimes justifiable over safety) and expensive capital outlay on
generating plant.

   Coal cutting machines were introduced only very slowly into British
mines whereas in American mines the opposite was the case. In 1913 only
8 per cent of the total British output was mechanically cut and this figure
had only risen to 14 per cent in 1921 and 31 per cent in 1930. In America

42 Hewers digging coal by hand at a longwall face in
a British mine in about 1910. Contrast this with
illustrations of machinery in modern mines

60 per cent of the coal was mechanically cut in 1920 and the figure had
risen to 78 per cent in 1929. Britain had many small mining concerns
and the necessary capital was not available for the purchase of expensive machines. In addition the seams in British mines were often un-
suitable for easy use of machinery since they were frequently shallow and
irregular.

Developments in underground mining technique were not confined
to coal cutting machines. The movement of coal underground from the
face to the shaft showed equally interesting innovations. In the early
nineteenth century many mines introduced wheeled tubs for this process.
Filling the tubs was done by hand but in seams which were too thin to
allow the tubs to be taken along the face to be filled, a great deal of extra
work had to be done to move the coal to the road head where the wagonway
terminated. If a seam was 2 foot thick and with walls 15 yards long the coal
had to be thrown from the extreme end three or four times before it
reached the road head, resulting in much effort and much broken coal. The
answer to this was a coal conveyor, and the earliest conveyor consisted of a
long shallow wagon mounted on rails and moved backwards and forwards
along the face by means of an endless wire rope, fastened to each end of the
wagon, and passing round a wheel 10 inches in diameter fixed at the far

end of the face, and also round another, the driving wheel, fixed at the road head. To one of the spokes of the driving wheel was fixed a handle about 12 inches long which was turned by a boy when the wagon needed to be moved along the face. The Blackett conveyor, the first successful conveyor introduced into British mines, was far ahead of the one described since it was driven by compressed air and consisted of a continuously moving shallow trough which carried the coal to the wagons at the wagonway. As in the case of coal cutting machinery, the American mining companies were more ready to introduce coal conveyors than their British counterparts.

Moving the wagons to the shaft was done in a variety of ways. In many British mines the loaded wagons were pushed by the putters or sometimes they were pulled by small pit ponies to the main wagonways. Once they reached the main wagonway they were hauled to the shaft by a stationary steam engine. As early as 1805 the British engineer John Curr claimed to have pioneered this idea, no doubt being inspired by the developments of steam power on the surface. The most typical system in use from the mid-century onwards was the endless rope system. The rope passed round a driving pulley on the mine bottom, along the main wagonway, and round a pulley at the opposite terminus. This system, powered by either a compressed air motor or later by an electric motor, travelled continuously in one direction. The wagons were hitched to the rope by means of hooks or clamps and hauled in and out as required. Before the First World War small compressed air locomotives were introduced to haul wagons along the main underground roadways, particularly in American mines. Electric locomotives taking their current from overhead wires in exactly the same way as street cars were introduced between the wars. More recently diesel locomotives have been used.

In the late nineteenth century much was done to make the winding of men safer and the winding of coal more efficient. In British mines almost all winding was done by steam engines; electrical winding before the 1920s was very exceptional. No large-scale electrical winder had been built before 1905, but after this date they were widely used in France, Germany, and America. As late as 1924 coal was electrically wound from only 182 British mine shafts.

Slow progress was made in the introduction of electric lighting into mines. By 1912 more than half of British mines used no electricity to speak of; the introduction of electric lighting was hindered by the comparatively high cost of power before 1914. By this date the more progressive

43 *Left:* original Davy safety lamps. These small oil lamps encased in a wire gauze cylinder gave a safe light and warned of the presence of gas by the changing colour of the flame. *Right:* naked lights such as this caused serious accidents

mines used electric lighting for shaft bottoms and main wagonways. The miner at the face still used a lamp based on the pattern of Sir Humphry Davy's oil lamp introduced earlier in the nineteenth century. Portable electric lamps were rather cumbersome and although they gave a more powerful light than the Davy lamp they did not give any indication of gas in the workings. There were mines in many parts of the world where gas was almost unknown and where naked lights were used. The whole problem of mining safety became more acute when in the last decade or so of the century it was shown beyond a shadow of a doubt that coal dust was responsible for many explosions, either by itself or when mixed with firedamp. Almost all mines are dusty except the shallowest and a few exceptionally wet mines. Much dust is created in the cutting, loading, and hauling of coal to the shaft and it is blown through mine workings by the ventilation system. When the dust in the air is sufficient in quantity, any large hot flame will ignite it and cause an explosion. Serious explosions occurred in British and French mines (especially at Courrières) which were free from firedamp and where naked lights were used. Only slowly were the enormous dangers present in a 'dusty' mine recognized. The greatest mine disaster ever recorded, at Honkeiko in north China in 1941, involving

44 An electrically driven continuous mining machine operating in an American mine. The Americans pioneered the use of such machinery in the late 1940s

the loss of over 1,000 lives, was caused by such conditions, whilst the Senghenydd disaster in Glamorgan, Wales, on 14 October 1913, which led to the loss of 439 men and boys, was due to a number of factors but the ignition of coal dust was one of them. Coal mining in Europe and America exacted a heavy toll in lives earlier in this century, although contrary to expectations the highest loss of life was not caused by explosions but by a multitude of other accidents.

Since the beginning of this century, when coal mining was gradually becoming mechanized, rapid strides have been taken in the development of mining machinery. Drilling, blasting, and hand loading on to conveyor belts made up the routine pattern of coal mining. Mines using the longwall method followed what was known as the 'cyclic' system of mining. Coal cutting usually took place on the morning shift. The coal was undercut, props and bars were reset, and shot holes were drilled in the face. At the end of the shift the shotfirers inserted charges and blasted down the coal. The second and main production shift loaded the coal and moved it out of the mine. The final shift built stone packs to support the roof behind the face. Roadways were advanced to match the advance in the face, and their roofs made secure. The belt conveyor along the face had to be dismantled,

45 A power loader cutting and loading coal on to
a conveyor belt, with self-advancing hydraulic
supports on the left

manœuvred along the forest of props and reassembled in a path closer to
the face. Therefore only one shift in three was productive. Engineers were
anxious to develop machinery that would make it possible to produce coal
on all three shifts, the system known as 'continuous mining'. The development of a machine that cut the coal and loaded it in one action, the cutter-
loader, made possible two productive shifts, not just one. The third shift
moved the cutter-loader forward and secured the roof. The final development was the introduction of a machine which cut the coal, discharged it
into a conveyor, both cutter and conveyor advancing forward once a
section of coal was cut. This was dependent upon the development of a
new type of roof support that could hold up the roof as the cutter conveyor
moved forward. The hydraulic prop was the earliest such support – a
telescopic prop that gave rigid support to the roof when required but was
withdrawn and reset easily by operating a valve, when the face advanced.
These hydraulic supports led to the introduction of self-advancing supports
which, at the operation of valves, advanced automatically to support the

46   Mining by remote control. The control room
in a Yorkshire mine for the automatic coal
conveyor system

newly exposed roof as the face advanced. Technical advances such as these
rendered cyclic mining obsolete and led to the introduction of continuous
mining in which every shift was productive. Continuous mining began in
America in the late 1940s and in Britain and Europe in the early 1950s.
American mines use the pillar and stall method of mining whereas most
British mines use the longwall, and hence there are differences in the
coal-face machinery used, but the principle is the same, that coal should be
extracted on all shifts, not just one in three. However, it must not be
assumed that all coal mining is done in this way. In some British, German,
and Russian mines mechanical picks are used to loosen the coal and it is
still hand shovelled to a conveyor belt. The Donbass coalfield in the
Russian Ukraine is not worked by the continuous mining process because
the seams are not very accessible and working conditions are difficult. But
the coal is high grade and therefore worth mining. Clearly a balance is
achieved between machines and men depending on conditions.

The logical development from continuous mining was mining by remote

85

control. In Britain remote control mining was begun to exploit seams of good-quality coal where it was difficult for the miner to gain access. The result has been two pilot schemes described as 'Remotely Operated Longwall Faces', usually known as ROLF, the first experiment in the world of mining by remote control. ROLF 1 came into operation in January 1963 at Bevercotes colliery, Nottinghamshire, and ROLF 2 a year later. The two basic pieces of equipment for mining by remote control were the power-loader and the self-advancing hydraulic roof supports; the former was a machine which both cut coal from the seam and loaded it mechanically on to the face conveyor whilst the hydraulic roof supports advanced automatically to support newly exposed roof as the face advanced. All operations were controlled by one man, seated at a control console located in the roadway at the end of the coal face which carried the coal away towards the shaft. This console was part of an array of equipment mounted astride the coal conveyor, the coal passing beneath. Between the man at the console and a small staff at the shaft bottom, the only human workers were the maintenance patrols. Results from these pilot schemes were encouraging. Their productivity was higher than that of manually controlled faces with similar equipment. In one week ROLF 2, operating under full production conditions, attained an output of 23 tons a manshift, compared with the national average for mechanized faces of $6\frac{1}{2}$ tons a manshift.

Developments in surface mining have been equally spectacular with the introduction of strip or open cast mining, a technique made possible by the development of giant shovels that can remove overlying soil and rock, known as 'overburden', to a depth of 120 feet. It must be remembered that open cast mining was probably the first type of coal mining undertaken by man, it has long been the method for mining brown coal or lignite in Europe, and in metal mining the method is very common. Somewhere about 1886 open cast mining began in Illinois by a simple method of loosening the surface with a plough, scraping away the soil, and exposing the coal which was removed by pick and shovel. Machinery was introduced in the first years of this century when steam shovels capable of lifting 8 tons of coal were used, the overburden being broken up by blasting if necessary. Early in the nineteenth century lignite was mined in much the same way as in Illinois in the 1880s, although throughout most of the century much lignite was mined underground, only 20 per cent being removed by the open cast method. The lignite open cast mines of West Germany are a

47 Lignite, or brown coal, being mined in Poland
with a machine that digs the coal and moves it along
a conveyor belt to waiting trucks all in one operation

48 Open cast coal mining in Illinois, where a 2,100 ton
wheel excavator digs off overburden on the left of
the picture and dumps it on to a spoil bank on the right

good example of the main features of such mining done on a large scale. The first stage is to remove the overburden with large excavating machines which scoop it up by means of chains of buckets. Waste material is carried away and tipped into exhausted workings. Coal is worked on a single long face, up to 2 miles long and 300 feet high according to the thickness of the seam. Seams of 30 to 60 feet are common while over 300 feet is known in places. On the coal face a wide step or bench is cut, on which run the machines which work the coal. These either scoop up the coal from below in chains of buckets or claw it down from above, so that the whole face, with the bench, gradually advances. In certain parts in America where physical conditions permit, large quantities of coal are mined by such methods. In 1965, 31 per cent of all coal mined in America was obtained in this way.

In the last ten to fifteen years American engineers have developed enormously powerful machines to mine by the open cast method. The wheel excavator first used at the Red Ember coal mine in Illinois performed three functions simultaneously; it dug the overburden, transferred the material to an endless belt, which deposited the material on top of a spoil bank. The coal exposed by the excavator was blasted loose with dynamite and electrically powered shovels scooped up the broken coal into wagons. This excavator weighed just less than 1,250 tons and consisted of a digging apparatus, a conveyor system to carry the freshly dug material to the spoil heap, and a large self-propelled chassis on which the first two parts were mounted. The digging wheel was the unique feature of the machine; it had a diameter of 24 feet and was equipped with nine buckets. It turned at a speed of six to seven revolutions per minute. To carry the sixty-three bucket loads per minute the first part of the conveyor system travelled at the rate of 1,070 feet per minute and from this belt the material flowed on to the stacker belt travelling at 1,170 feet per minute. The total conveyor system length was 355 feet. During one year of operation this machine averaged a removal of 1,500 cubic yards per hour of operation. Since the Red Ember coal mine used the first wheel excavator in the late 1950s American engineers have developed much more powerful ones, and such a machine at Cuba, Illinois, can remove 3,500 tons of overburden an hour. The output from such mines is enormous; the Red Ember mine in 1961 could mine up to 7,500 tons of prepared coal per day.

Open cast coal mining was not introduced into Britain until 1942 and is unlikely to reach the scale of the American industry. Large reserves of

88

49 Augur mining in the Appalachians showing the
drill, or augur, extracting coal from a seam and
loading it on to lorries. More augur sections for
deeper drilling lie beside the machine.

coal do not lie immediately below the surface in Britain as in some American states and in general conditions do not favour this type of mining. Only about 7 million tons of coal are mined by the open cast method in Britain.

Since 1945 American mining engineers have developed auger mining, a natural development from open cast mining. In hilly country open cast machines followed the contours and stopped when the seam disappeared into the hill since the cost of stripping the overburden was high. On such sites giant recovery devices called 'augers' are now used which can bore up to 200 feet into a seam in a hillside and bring out cores of coal to a maximum of 5 feet in diameter and 200 feet in length. Only a small percentage of coal is mined in this way, mostly in the northern and mid-Appalachian fields.

Mining life, especially for those underground, has changed enormously because of the many innovations since the mid-nineteenth century. The miner's job has become easier and much safer, but the safety factor did not

50   Houses in close proximity to a mine have always
been typical of mining communities; this one in South
Wales shows something of the life of the coal miner

immediately follow the introduction of machinery. The late nineteenth
century and the early twentieth century were characterized by high
accident rates and violent struggles between managements and miners for
better wages and conditions. Average death rates per year in mines over
the period 1897–1911 were as follows:

America	3·31 per 1,000
Germany	2·21
France	1·52
Britain	1·32
Belgium	1·03[2]

The high death figure for America was due to the increase in deep mining,
the growing use of electricity for haulage work underground and machine
coal cutting, but also because there was little regard for human life. After
the First World War the number of deaths per thousand in American mines

[2] H. Stanley Jeavons, *op. cit.*

was three times as high as in Britain. An American Coal Commission made a thorough investigation in 1922–23 but its recommendations for greater safety regulations were largely ignored. Equally serious were the struggles to raise wages, especially in the inter-war years when the coal industry languished in the Western capitalist world, and coal owners faced with declining profits were met by insistent demands for wage increases. The outcome in Britain was the General Strike of 1926 from which miners emerged worse off than ever, whilst in America labour relations were even more embittered. During a national coal strike in 1922 twenty strike breakers were killed after violent fighting with striking miners in Williamson County, Illinois. Generally speaking the lot of the miner did not improve until the Second World War when demand for coal increased. For example, wages of British miners rose steadily after the war and underground workers were considered to be among the élite of manual workers in the early 1950s. Side by side with such developments has been the provision of welfare benefits such as baths for miners at the end of shifts, slow to be offered in Britain since they were first introduced in 1902, but compulsory in Germany from an early date in this century.

Since the end of the Second World War, and particularly in the 1960s, redundancy has faced many miners. Coal as a source of heat and energy has been challenged by other fuels, particularly oil and natural gas. In countries such as America, Britain, Belgium, and Germany the demand for coal has dropped. Numbers employed in coal mining have been reduced even more because mechanization has greatly increased the productivity per man per day. The challenge from new fuels has led to the closure of many uneconomic mines and a further decline in the labour force. In American mines, average productivity per man in all mines more than doubled from 1940 to 1963, whereas the number of workers decreased by more than 60 per cent. In the countries of the Coal and Steel Community 100 mines were closed from 1958 to 1963 and the number of underground workers decreased by nearly 25 per cent.

Changes such as these have meant there has been an urgent need to provide alternative employment for redundant miners. In areas where the bulk of the labour force relied on mining and there was little diversification of industry, this has not been easy to provide and it has led to chronic social problems. These have been very acute in areas such as west Kentucky in America and the Northumberland and Durham coalfield in Britain. Even if alternative employment has been available some distance from their homes, the miners have been reluctant to move to these areas. This is largely because of the nature of coal-mining communities. They tend to be small and close knit; mining families show a marked reluctance to move to an area that they do not know from a community with which they are familiar. Some miners, especially those in their fifties, are often too old for re-employment and remain permanently unemployed. Social problems such as these are not to be found everywhere since some countries, notably Russia, have an expanding industry where new fields are being developed and there is an urgent need for miners. Coalfields such as the Donbass and the Kuzbass have steadily rising production figures which could well indicate an expanding labour force. The Pechora field in north east Russia near the island of Novaya Zemlya used the forced labour of Polish and German war prisoners to develop the field in its initial stages.

Finally, it is worth examining the future prospects for the coal industry. Until the mid-twentieth century coal was a major source of energy in all the important industrial countries of the world; for example in America in 1900 coal provided 89 per cent of all the energy needed. In a number of countries the heavy dependence on coal has been considerably reduced; for example it provided only 22 per cent of all the energy needed in America in 1961. Thus the inevitable question arises, has 'King Coal' been dethroned and is the end of the coal industry in sight? In countries such as Germany, America, and Britain the challenge from oil has been felt very acutely. During the 1960s deposits of natural gas discovered in the North Sea have offered a further challenge to coal as a major source of fuel and energy in western Europe. Potentially the greatest challenge comes from nuclear power, a major new source of energy. Since 1945 scientists have pushed ahead with the development of nuclear power from uranium as a source of heat and energy. By the early 1960s there were seventeen nuclear power plants either under construction or operating in America whilst in Britain the first nuclear energy was fed into the national grid in 1960. Many people thought that the development of nuclear power would lead to a drastic reduction in the demand for coal; this has not happened, due to the high capital cost of developing a nuclear power station.

Has coal been dethroned? The answer to this question depends on where one is and how one looks at things. In America it has ceased to be the major source of energy and its position is seriously challenged in countries such as Britain, Belgium, and Germany. Yet, total world production of coal has continued to rise and in some countries coal production has increased

51 Natural gas obtained from drilling rigs like
this one in the North Sea offers a serious challenge
to the coal industry in the 1970s

considerably since the Second World War. Russia is a good example of such
a development since she has increased her production of coal from 166
million metric tons in 1940 to 600 million metric tons in 1965. China has
large reserves and whilst it is difficult to discover exact data it seems that
since 1949 her production has increased rapidly, producing about 22 per
cent of total world output of all types of coal in 1962. Coal is still of
enormous importance as a source of energy and a source of valuable by-
products. Economic geographers suggest that world trends show that coal
reserves are destined to outlast those of gas and oil. 'King Coal's' crown
may have slipped and appear somewhat tarnished but we are a long way
from the revolution that will overthrow him for good – yet one day all
world stocks of coal will be exhausted.

7 Diamond mining

UNTIL THE EIGHTEENTH CENTURY the chief source of diamonds was India, particularly the states of Madras, Orissa, and West Bengal. The town of Golconda was famous as the market for the diamonds of the country but in the early eighteenth century India's position as the leading diamond producer was challenged by Brazil. In 1725 diamonds were discovered at Tejuco (now Diamantina) and by 1740 the mining of them had become an important industry; the Bahia mine in particular was famous for the quality of the stones it produced. But within four or five years after the discovery in 1867 of diamonds in South Africa, such was the quality and quantity of the stones, all other known deposits in the world shrank into insignificance.

The first discoveries were made near Hopetown, a small agricultural settlement on the Orange River, some 30 miles from its junction with the Vaal. Initially the discoveries attracted little attention; not until 1869 could it be claimed that a rush had begun. The earliest mining centre to develop was the town of Pneil; around this town, largely on the banks of the Vaal, miners washed river gravel in order to locate diamonds. Diamonds are found particularly in volcanic Kimberlite pipes and fissures, and in alluvial terraces. The earliest discoveries were made in gravel on these alluvial terraces along the Vaal and Orange rivers. The machines for washing the gravels were of two kinds, cradles and cylinders, and very similar to those used by the earliest miners in California. The principle of both types was the same; the gravel was passed through two or three wire sieves of differently graded mesh in the machines, so that whatever stones the sand and gravel held were caught in the meshes, while water was poured over them, washing the sand and gravel away. The cradle was mounted on rockers, and like the Californian cradle had a long handle by which it was violently rocked during washing. The cylinder was revolved, the result being the same as in the cradle, the sand and gravel disintegrated and then the contents of the sieves could be examined for diamonds, the waste thrown away, and the process begun again.

52 This simple cradle device used by diamond diggers in the Kimberley area in the 1960s was probably similar to that used by the earliest diamond miners in the 1870s

This mining area lasted only a short time because in September 1870 diamonds were discovered on Dorsfontein farm at Dutoitspan, some 20 miles south east of the alluvial diggings. This area yielded diamonds without washing and the river diggers of the Pneil area, tired of standing knee deep in water whilst cradling, moved to the new field leaving behind them in the words of one digger '. . . nothing much but empty sardine and canned meat tins, paper collars, broken pipes and mud walls'.

These dry diggings attracted much attention in South Africa but by July 1871 three more areas very close to Dutoitspan had been discovered where diamonds were found in considerable quantities: Bulfontein farm, Vooruitzigt farm known as the De Beers mine, and on Colesberg kopje (hillock) known as New Rush, but from 1873, as Kimberley. These discoveries

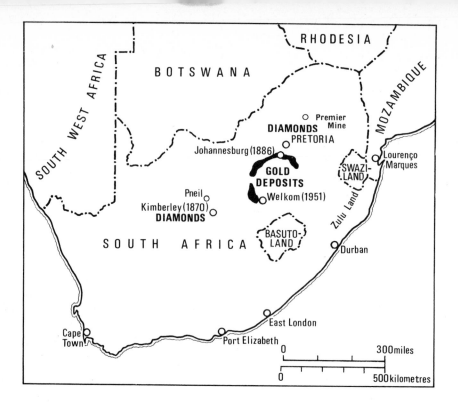

MAP NO. 4 GOLD AND DIAMOND MINING IN SOUTH AFRICA IN THE NINETEENTH
AND TWENTIETH CENTURIES

attracted world-wide attention for soon a rush began similar to the gold
rushes of Australia and America. The De Beers mine lay so close to Kimberley that it became part of the new boom town that developed, whilst
Dutoitspan and Bulfontein could be best described as suburbs of the new
town that was soon second only to Cape Town in size. This new town was
in a desolate area, with only primitive conditions and few resources to
support a population that quickly grew to about 20,000, of whom just over
half were non-white. This description of the miners' suburb of Dutoitspan
shows that the town had all the features of a typical mining camp set in the
frontier of an expanding area:

> Canvas shelters everywhere, and as the coach got into one of the roads
> leading directly to Dutoitspan, the only wooden buildings seen were made of
> packing cases, though dismal looking iron shanties intermingled with mud
> heaps, wells and washing apparatus were on view by the score. However, on
> we went; tents got more numerous. Naked Kaffirs and dogs appeared in
> plenty . . . The driver, livening up the tired mules, showed his gleaming teeth

96

and looking supremely happy, drove up the Main Street of Dutoitspan in what he considered fine style . . . Although it was the principal thoroughfare, many small bell and wood-framed canvas tents lined either side of the Regent Street of Dutoitspan.[1]

The earliest miners believed that the diamonds were to be found solely in the gravels that lay on the surface in this area but soon they found that they were restricted to vertical pipes having a roughly oval cross section and nearly vertical sides. In the area of Kimberley four of these pipes were found; they were filled with a heavy dark blue rock called 'blue ground' or 'Kimberlite'. The miners were allowed to peg out claims 30 feet square. The actual mining at this time and for some years was carried on in a most individualistic, uncooperative, and amateurish way; in all mines the method employed was similar. The ground was broken with pickaxe and shovel and then raised to the surface by African labour or by carts or barrows that were driven or trundled to the sieves. Rarely were tunnels dug, the diggers merely deepened their hole. Many miners employed African labour to carry the sacks or buckets up knotted ropes hanging over the pit, or by small niches cut in the rock, or by ladders made of rope or of wood. In fact raising the broken rock to the surface was the most serious problem. By 1872 many miners had adopted the device of two grooved wheels, one fixed at the pit bottom, the other at the surface, with a rope to which two buckets were attached passing over the wheels. On turning the wheels by the handles attached to each, a full bucket could be raised and an empty one lowered. Eventually by 1874 pits became so deep that horse whims were introduced on many claims. By this time the 15 feet wide roadways which had existed between claims had disappeared, and the Kimberley mine in particular was a spectacular sight, with innumerable wire ropes stretching from all points at the bottom to all points at the top, some 1,000 feet long. Trollope, the novelist, visiting Kimberley in 1877 wrote this about the mine:

> . . . you look down into a huge hole. This is the Kimberley mine. You immediately feel that it is the largest and most complete hole ever made by human agency . . .
> You stand on the marge and there, suddenly beneath your feet lies the entirety of Kimberley mine, so open, so manifest and so uncovered that if your eyes were good enough you might examine the separate operations of each of the three or four thousand human beings who are at work there. It looks so steep that there can be no way to the bottom other than by aerial contrivances

[1] Quoted in Oswald Doughty, *Early Diamond Days*

53 The Kimberley mine in early 1872, showing
the varying depths of the claims and the
roadways between them

> ... It is as though you were looking into a vast bowl, the sides of which are
> smooth as should be the sides of a bowl, while round the bottom are various
> marvellous incrustations among which ants are working with all the usual
> energy of the ant-tribe.

When Trollope went down into the pit he quickly discovered the difficulties
of clambering from claim to claim, since they were of varying depths, the
discomforts of heat and thirst, plus the dangerous conditions under which
the miners worked. His reservations were well summarized when he wrote,
'The going up and down is hard, everything is dirty, and the place below is
not nearly so interesting as it is above.'

In the four great mines, the Kimberley, the De Beers, the Bulfontein,
and the Dutoitspan, there were over 3,200 full claims, many of them sub-
divided. The problem of raising the rock and earth from the mines was
acute, and the change in 1874 from manpowered to horsepowered windlass
was indicative of the not far distant day when steam winding engines would
be introduced. Such machinery needed capital not always available to the

54 The Kimberley mine in 1877; the roadways between the claims have vanished and the scene is one of apparent confusion, with hundreds of claims being worked at different levels and a vast cobweb of haulage ropes

individual miner. The miners' problems were increased by the water that accumulated in the bottom of the mines and which compelled collective action and the use of machinery. When depths of 400 feet or more were reached, mining became dangerous since rock falls were frequent – burying claims, killing, and injuring many miners.

The problems created by a large number of claims being worked in a comparatively small area could only be solved by amalgamation and mining being taken over by companies. Before the big companies moved in a steady process of amalgamation took place among the miners, the uneconomic moving in with the economic. By 1880 there were almost seventy companies operating on this diamond field, the most important being run by Cecil Rhodes' De Beers Mines and Barney Barnato's Kimberley Mining Company. Rhodes, the son of an English country parson, had a tremendous struggle with the London East End Jew, Barnett Isaacs, popularly known as Barney Barnato, for control of the mines in the Kimberley area.

By 1888 Rhodes' company had outmanœuvred Barnato's company and Jules Porge's French company so that by 1890 De Beers Consolidated controlled 90 per cent of all the diamonds mined in the Kimberley area. In the same year Rhodes obtained control of the Premier Mine in the Transvaal and his gigantic company's domination of all diamond mining was completed.

These changes meant that to all intents and purposes the day of the individual miner was over, but it was not Rhodes's achievement to eliminate him; water, falling rock, rising costs had done that. The companies in the late 1870s had been compelled to introduce new methods in view of the problems. Shafts were sunk some distance from the rim of the pipe and tunnels were driven from these shafts into the kimberlite rock. The main hoisting shaft was usually sunk to a depth of 600 to 800 feet, well below the lowest levels worked. Working levels were about 40 feet apart vertically. The mining skill was provided by white men; the labouring was done by Africans. This division had existed before the company era and was not initially due to colour prejudice. The Africans who came to the fields had no skill to sell, their stay was often brief, and they returned to their kraals once their short contracts expired. The white miner was usually a man of experience in mining and invariably not of South African birth. South Africa simply had no skilled workers on whom the mines could draw. The white miner took good care that the African was rarely allowed any position that challenged his monopoly of skilled jobs. In the early days on the diamond field the principle was established that skill and high wages were a privilege of the white race, while the heavy labour and menial tasks were the province of African labourers; a principle followed in the gold mining industry once it developed in South Africa and faithfully followed ever since in mining in the republic.

The contribution of the diamond industry to the early development of South Africa was considerable.

55 [opposite] The first horse whim used at Kimberley mine in 1874 to haul materials to the surface. Similar machines were used in European mines until the early nineteenth century. 56 An African mining crew supervised by a white man prepare to blast down a section of the diamond-bearing 'blue ground' in the Premier Mine, Transvaal

57 Beach mining for diamonds in South West
Africa. Movable interlocking concrete blocks, with
specially designed cavities to dissipate the force of
the waves, stand behind the plastic-covered sand
supporting wall.

Diamond mining carried forward what wool had begun but had been
unable to carry to completion. It provided a greater incentive and more
substantial means for the modernization of the lumbering transport system. It
attracted population and capital to the country, diversified the life and broad-
ened the opportunities of young men and women, and gave strength and
purpose to political life.[2]

The pace of change was increased again by the gold mining industry
which developed from 1886, an industry which relied heavily on the capital,
skill, and expertise of the diamond magnates for its early and successful
development.

The De Beers company has continued to dominate the South African
diamond mining industry in this century. The Kimberley mine was worked
from the surface to a depth of 3,250 feet until 1914; since then it has been
mined from underground. The bulk of South African production has come
from four mines in Kimberley, two in the Orange Free State, and from the
largest of them all, the Premier Mine near Pretoria in the Transvaal. This

[2] C. W. De Kiewet, *A History of South Africa, Social and Economic*

58   A more detailed view of beach mining, showing
a machine scooping up gravel in the background
while miners clear the remainder by hand to expose
bed rock

mine covered an area of 78 acres on the surface; it is now mined from below ground and has been worked to a depth of over 3,500 feet. In 1905 the Cullinan Diamond was found here; it weighed 3,025 carats, or approximately $1\frac{1}{3}$ lb, and was presented to Edward VII of England by the Transvaal Government in 1907. It was sent to Amsterdam to be cut and the nine gems it produced are now part of the English Crown Jewels.

Diamonds are mined underground by the block caving technique using the degree of mechanization typical of metal mines. The mines rely heavily on African labour, an important factor in the continuing success of a number of them since the average South African mine moves 30 million parts of ground to secure one part diamond.

In 1927–28 coastal deposits of fine gem diamonds were discovered along the Atlantic coast, both south and north of the mouth of the Orange River. Scientific prospecting revealed three ancient elevated beaches along the coast, containing diamond-bearing deposits varying from a few inches to more than 20 feet in depth, with an overburden of sand varying from a few inches to 40 feet in thickness. These deposits are worked as open cast mines where a great deal of expensive earth-moving machinery is used to strip the overburden and remove the diamond-bearing deposits.

The most interesting development in diamond mining has occurred in South West Africa, which is administered by the South African Republic, where specially equipped ships are reclaiming diamonds from the sea-bed. The first vessel to carry out this work was launched in April 1963 and by July had proved the existence of diamond-bearing gravels on the sea-bed in the Chameis Bay area, about 60 miles north of the mouth of the Orange River. The ship was capable of treating 18,000 tons of gravel a month and early results were highly encouraging. The first such vessel was wrecked in a storm in July 1963 but the recovery of diamonds has continued, using improved methods.

Gem diamonds are only shown to their best advantage if they are skilfully cut. Once this technique had been perfected by the seventeenth century the demand for diamonds began to increase and from this time gem stones were regarded as being of great value. Diamonds such as the Koh-i-nor, which was probably mined in the early fourteenth century, were eagerly sought and gem stones continue to excite admiration. People find it advantageous to invest their money in diamonds, since their market value remains high and investors are protected from the violent fluctuations in the value of paper money. The film actress Marilyn Monroe might have summarized the female attitude to gems with her song 'Diamonds are a girl's best friend' but industrial diamonds play an important but less glamourous role. They now form 80 per cent of the total world diamond production. They are used in modern industrial drilling equipment where their hardness is of special value; from oil-well drills to dental drills diamonds have their use. Bort, the cheapest grade of industrial diamond, is crushed and graded into powders for a great variety of grading and polishing operations. The chief use of diamond powder is in the manufacture of grinding wheels for sharpening metal cutting tools. A more recent use of diamonds has been in automatic machinery where diamond-bearing parts are valuable because they wear longer, and there is less frequent need to stop the machines to replace worn parts.

African territories, particularly the Congo and Ghana, are important for the quality of the industrial diamonds they produce, the Republic of South Africa having been famous for the quality of the gem stones it has mined. Because of the increased demand industrial diamonds are no longer a by-product of gem stones and are now mined for their own sake. Since

59 Not all alluvial diamond mining in South
Africa is done with sophisticated equipment, as this
variety of fairly old machinery indicates

1929 the annual output of industrial diamonds has exceeded that of gem
diamonds, the Congo and Ghana having played a part in this enormous
increase.

All diamond deposits in West Africa are alluvial. In Ghana, where deposits lie on river beds, diamonds are recovered by using dredges similar to
those used in California to extract gold from alluvial deposits. In some
areas the diamond-bearing gravels are exposed by stripping off the overburden with gigantic shovels; the gravel is then dug out and taken to the

treatment plant. In West Africa, but particularly in Sierra Leone, large quantities of diamonds lie close to the surface and can be recovered easily using primitive tools. This has attracted large numbers of Africans who mine diamonds illegally. As early as 1952 this primitive surface mining, or 'potholing' as it is known, began and by October 1956 some 75,000 Africans were involved. Strenuous efforts by the police have reduced their numbers but the *Daily Telegraph* reported on 2 January 1970 that illegal mining was carried on by something like 10,000 Africans. The government of Sierra Leone is losing heavily since it imposes a tax on all exported diamonds and obviously it does not collect this from the illicit miners. Potholing for diamonds is serious because it recovers only one-quarter to a half of the stones, making the deposit uneconomic for further working. However, in spite of strenuous efforts by the Sierra Leone government, the practice continues.

It has been claimed that a deficiency of diamonds would cause a breakdown in modern metal-working industry. This perhaps explains in part attempts that were made to produce synthetic diamonds. In 1906 Sir William Crooks succeeded in making an artificial diamond but it was of no commercial value. However, the General Electric Company of the United States claimed in October 1957 that it had succeeded in making an artificial diamond of commercial value. Its claim was justified and since then artificial diamonds have been manufactured in America and South Africa, and the Soviet Union has been making diamonds from some time before 1957. As yet the artificial diamond has not been able to offer a serious challenge to the natural diamond but, with the ever increasing exhaustion of diamond deposits, the natural diamond may one day be ousted.

8 Gold mining in the twentieth century

THE GOLD DISCOVERIES of the nineteenth century were usually the work of individual prospectors or very small groups and led to spectacular gold rushes. The twentieth century has seen occasional gold strikes by individuals but in general the bulk of the alluvial gold that can be easily worked is exhausted. The discoveries of the Witwatersrand goldfields of the Transvaal in 1886 pointed the way to the future development of the gold mining industry, not only in South Africa but in the rest of the world. There was little alluvial gold in the Transvaal: there were no glittering Australian nuggets, such as that discovered at Ballarat in 1853 weighing 132 lb 8 oz, or laden Comstock lode. The Witwatersrand ores were located in reefs buried deeply in the earth. Fortunes could not be made unless a person had money in quantity to invest in the development of a mine. In America, Australia, and the Yukon there was a gap of some years between the development of shaft mining and the exhaustion of the alluvial gold; this gap was not present in South Africa. From the early 1890s the gold mining industry of that country has been dominated by the large businesses. This pattern has continued in the twentieth century in most gold mining countries but it occurred first in South Africa in the last decade of the nineteenth century. In this aspect the history of South African gold mining belongs to a chapter on the twentieth century rather than to one on the nineteenth.

The discovery of the Witwatersrand goldfields was not accidental. The mineral resources of the Transvaal had been known for some time; as early as the 1850s the Boer farmers seem to have been aware of them. The spectacular gold rushes in America and Australia had directed attention to South Africa and in 1886 a man called George Harrison discovered gold, although the first recorded discovery of gold in the Transvaal was made by one Carl Mauch near the Olifants river in 1868. In July 1886 the first sample of ore was taken to Kimberley and tested. Its richness prompted a speculator to travel straight up to the Witwatersrand and buy the farm on which the gold was found for £7,000. Within a few months of this discovery mining began on an important scale. The rapid development of the gold-field owed much to local conditions. Not far from the Witwatersrand was Kimberley, already famous for its diamond mines and an invaluable source of money and experience of mining conditions. The machines essential for deep mining needed steam power and although there was little wood and no water power, there was plenty of coal. The abundance of cheap, good-quality coal was of utmost importance to the gold mining industry. Of equal significance was the discovery and successful application of a new process of extracting gold from the ore known as the 'cyaniding process'. There was little doubt that the Witwatersrand deposits were very extensive but the average gold content per ton was low, lower than in any other gold-field in the world. Thus the problem was how to make the poor ore pay.

Until 1889 (the date of the discovery of the cyaniding process) there were two principal methods of extracting gold from crushed rock; the use of mercury and the use of chlorine gas. Mercury had been used by the Spaniards in America and the process by the mid-nineteenth century involved passing the crushed rock over copper surfaces which had been amalgamated with mercury. The particles of gold stuck to the amalgamated plates and could be recovered by evaporating away the mercury. Chlorine gas had become a commercial commodity during the middle years of the century. Chlorine was led over the crushed rock producing a soluble chloride of gold that could be dissolved out of the residue of the ore, and then be precipitated chemically from the solution. By these methods crushed ore carrying as little as one part gold in 30,000 was considered suitable for profitable working.

The greatest single technical advance in the process of gold extraction was the cyaniding process, discovered in 1889. Two Glasgow doctors, Robert and William Forrest, and a chemist, John McArthur, patented this process. The crushed ore was circulated through tanks containing a weak solution of cyanide, which has an affinity for gold. The solution dissolved the gold but not the rock particles which were filtered off. Zinc dust was added to the cyanide solution and it replaced the gold, causing fine specks of gold to be precipitated out. When this process was applied on a commercial basis to the Witwatersrand ore it was possible to extract 96 per cent of the gold from the crushed ore. Whilst there were initial difficulties in putting a method devised in the laboratory into practical operation, without any doubt this process saved the Transvaal boom. Science had turned into profitable ore what would have been barren rock to the Spaniards in Central and South

America in the sixteenth and seventeenth centuries, and indeed to the earliest miners of California and Alaska.

When gold was discovered in the Transvaal in the 1880s the republic was in a serious financial plight. The discovery of gold changed the situation completely, and the revenue rose from less than £250,000 in 1885–86 to £1½ million in 1889–90. By 1898 the value of the gold production had risen to £16 million. This vast increase of wealth brought serious problems too. Until the gold finds the Transvaal was a sparsely populated territory inhabited by Boer farmers who had successfully won independence from British rule in 1882 and whose pastoral way of life was slow and unchanging. Into this rural, unprogressive society moved the gold miners and the gold mining companies. The two did not mix; at once there was antipathy between the mining community and the farming community. The miners cared little for the Boers and their way of life and it seemed to the Boer farmers that they would be swamped by the newcomers, the 'Uitlanders' or 'Outsiders' as they called them. Their worst fears seemed justified; Johannesburg had grown from a mere village to a population of 80,000 by 1892 and 102,000 by 1896; by the late 1890s the Boers were outnumbered by 7 to 3. Kruger, the President of the Transvaal and the personification of all that the Boer stood for, took steps to bar the Uitlanders from all political influence. Their grievances were exploited by Rhodes who, as he had failed to force the Transvaal into a customs union for South Africa, now sought to foment dissident elements there. The outcome was the Jameson Raid at the end of 1895, a stupid piece of work which played a part in the causes of the Boer War which began in October 1899.

By the time the war had begun the goldfield based at Johannesburg was well established in spite of the high capital cost of starting a new mine. The gold-bearing reefs which outcropped near Johannesburg dipped underground at an average angle of 25° which meant that large companies were essential for the deep mines once the outcrops had been worked. In the 1890s it could cost up to $2 million to start a new mine. However, the earliest mines were little more than trenches that were dug into the reef and the gold-bearing ore was removed by hand by native workers. From the beginning large numbers of Africans were employed to do the manual work. Cheap African migrant labour was available in large quantities and proved to be an extremely important factor in the rapid development of the gold mines, although the supply of abundant cheap labour has led to slower mechanization of mining in the twentieth century. Nearly 100,000 Africans

were employed on the goldfield before the Boer War and their labour was important in a mining area where production costs were high from the beginning. Gold production was reduced as a result of the Boer War and to get the mines back into production afterwards Chinese labourers were imported. Some 50,000 came from 1904 onwards but they were repatriated by 1910, only after a considerable outcry about their living and working conditions.

Once the surface gold was exhausted new operations had to be undertaken. Vertical shafts had to be sunk to tap the reef as it dipped beneath the surface. At first the ore was drawn to the surface by machines similar to the whim gins in British and Belgian mines in the eighteenth century, the power supplied by bullocks. Then came the introduction of steam engines to replace animal power. Surface work was equally primitive; crushing the ore was done as the ancient Egyptians had done it, by raising and dropping heavy weights. The introduction of the stamp mill to perform this task quickly followed, the town of Johannesburg living with the incessant roar of these machines as they crushed the ore to powder for cyaniding. The replacement of hand drilling by machine drilling before the First World War increased output and efficiency. In the 1920s superior drills were introduced, in particular the jackhammer drill which was fitted with drills of better shape and improved steel able to bite into the hard quartz rock. The effect of such innovations was best seen in terms of production; in the first six months of 1914 the amount of rock crushed was about 250 tons per employee while in a similar period of time in 1930 the figure had risen to nearly 800 tons. Whilst production rose considerably there was a high accident rate underground of 4·64 per thousand in 1904 and a high death rate from silicosis of 1·9 per cent in 1927–28. One of the causes of such high rates was the carelessness in mining underground; high bonuses encouraged miners to take unwise risks. Round the clock working in some mines was a further cause, in some the underground air was never free from the dust of constant mining. Between 1912 and 1930 15,000 awards of compensation were made under the Miners Phthisis (Silicosis) Law. Since the 1920s greater care has been taken to lessen accidents and disease by single-shift

60 [opposite] A South African mine in the early 1920s, with a crew operating a machine drill. These replaced the hand drill but made mines dustier and noisier.   61   A South African drilling crew with a white foreman opening up a new tunnel. Water used to subdue dust whilst drilling takes place cascades to the floor from the machine (1960s)

operating and the introduction of air conditioning. The figures speak for themselves; the accident rate fell to 2·11 per thousand in 1934, and the incidence rate of new cases of silicosis fell to 0·80 per cent in 1936–37.

From its early days the goldfield in South Africa has been dominated by great mining finance houses. At one time there were nearly 450 companies, most of whom never paid dividends. Nineteen large companies paid dividends in 1889, only eight of them paid in 1890. Within two years of the discovery of gold four of the present seven finance houses had been established, all backed by men who had made their money in diamond mining. For example Gold Fields of South Africa was founded by Cecil Rhodes and Charles Rudd in the 1880s, the Barnato brothers founded the Johannesburg Consolidated Investment Company. These were the men who had money, could marshal the technical expertise needed underground, and were prepared to take the calculated risks necessary in the most hazardous area of gold mining in the world.

Since the 1890s there have been five major companies involved in the industry and only two new companies have managed to break into the select five; the Anglo American founded in 1917 and the Anglo-Transvaal Investment in 1933. The continuing success of these seven companies has been due to the improvements in geology and geophysics making it possible to gain a much more accurate picture in advance of the value of gold in the reefs. The success of the Gold Fields of South Africa finance company in the 1930s is a good example. In 1930 a Canadian-born engineer Guy Carleton Jones was made the company's Consulting Engineer at a time when rising costs and lower production led to a depression in the gold industry. Carleton Jones was persuaded by a German geologist Rudolf Krahmann that his company should use a newly developed magnetometer to try to locate the gold reefs that lay far below the surface. This instrument would not locate the gold reefs themselves but would pick up the pattern of the magnetic shales of the Lower Witwatersrand system. The gold reefs could then be easily charted since their position to these shales was already known. The magnetometer showed the pattern of reefs plus two hitherto unknown gold-bearing formations. With this information Carleton Jones persuaded his company to open up mines in a new goldfield revealed by Krahmann's work. This new field, the Far West Rand Goldfield, did not develop as rapidly as it might have done until after the Second World War because the gold lay buried very deeply.

Geology, geophysics, capital in enormous quantities, paved the way for the continuing development of the industry. Equally important was the flair and shrewdness of the late Sir Ernest Oppenheimer who made significant contributions to the gold mining industry until his death in 1957. Oppenheimer had founded the Anglo-American company in 1917 and shortly before the end of the Second World War he revitalized the industry by staking his own and his company's reputation on the potential of a new goldfield in the Orange Free State. Its existence had been proved by the geologists and geophysicists; Oppenheimer was the man who took a calculated risk. Results since have borne out his judgement as four of the seven most productive mines in South Africa are to be found in this field. His last gamble, the sinking of the Western Deep Levels in the Far West Rand Field, showed the enormous technical problems that had to be overcome to sink the mine and the high degree of skill and financial investment required in modern mining. Boreholes had shown that the rich Carbon Leader Reef (discovered by Krahmann) lay anything from 2 to $2\frac{1}{2}$ miles deep. A rock temperature of 130 °F was expected. Serious water problems were anticipated; it was expected that from each shaft something like 30 million gallons of water would need pumping. For almost fourteen years the sinking of the mine had been debated until finally just before Oppenheimer's death in 1957 the decision to go ahead was taken. It took just five years before the mine began producing. It cost $95 million to develop, yet in 1966 it made a working profit of $26·6 million.

Flair, business acumen, technical and scientific knowledge, and heavy capital investment have all played their part in the development of the gold mining industry of South Africa. Yet the enormous reservoir of cheap unskilled African labour has played a vital role in an industry that has always been faced with high operating costs. Basic wages for Africans are low but the gold companies have always argued that without the abundant supply of cheap labour the gold mines would have been priced out of business years ago.

The conditions under which the Africans work are probably unique, except perhaps for those of the gold miners in certain Russian goldfields. The Africans enjoy a kind of freedom hedged by considerable restrictions. The Chamber of Mines has two recruiting organizations which hire about 375,000 Africans a year, something like two-thirds coming from outside the South African Republic. Once they arrive at their assigned mine, having already been given fairly extensive medical checks, they sign a contract for nine months or a year. They receive a simple training in labouring tasks such as using a shovel, loading a truck, hauling ore out of the stope; whilst those with previous experience can graduate to more responsible tasks of driving underground trains, acting as foreman to a gang, or operating drilling machines. Since 1922 the African worker has not been allowed to move to any skilled work. A strike in that year arose over an attempt to allocate certain operations done by white workers to Africans and was only quelled by police and soldiers. From then onwards the African was kept in a subordinate position, although men such as Harry Oppenheimer urged that the African be given a greater degree of responsibility, and that certain jobs reserved for whites should be opened to Africans. In the 1960s some hesitant steps were taken in this direction but progress has been slow.

Throughout the period of contract Africans enjoy substantial welfare benefits. They are given free accommodation in a compound close to the mine. They sleep ten to twenty in a room, food is free and usually unlimited in quantity. Only 3 per cent of the total labour force are allowed to bring their families. Pay is low (£5 per month in 1970) but they are encouraged to save a portion of it for when they have completed their contract. Most compounds have a bar and recreation rooms such as a cinema. Free medical care is provided throughout their contract; the Anglo-American at Welkom in the Orange Free State have an 887 bed hospital, equipped with modern medical facilities, for the 50,000 Africans they employ in this goldfield. There can be no doubt that whilst the African is cared for during his contract period he has little real freedom, probably less than the average African who works in the large urban centres such as Durban. The work he does is arduous, dangerous, exhausting and, by most standards, badly paid. The conditions under which mining is done are vividly described by an English visitor to a gold mine:

> . . . shrouded in white overalls, enter a cage which plummets through a mile of rock in two minutes. There below is a noisy, hot, wet world lit by the dancing fireflies of the lamps on miners' helmets. A ten minute walk along a gallery cut through rock whose natural temperature is over 100 °F and any visitor is soaked by a combination of sweat and humidity. Then, above the constant hum of the air-conditioning and the rumble of trucks along steel rails, comes the sound of compressed air drills biting into solid rock. On one side of the tunnel a narrow opening begins plunging down at an angle of nearly 25 degrees towards the bowels of the earth. It is barely forty inches high and is delicately held open by props of blue gum. It is called, in mining parlance, a stope. Within the stope the rock seems to press in from all sides; tiny flakes fall from the roof into the pools of warm water in which everyone is kneeling or lying.

62 Washing gold-bearing gravel in the Belgian
Congo in the 1920s by fairly primitive methods
more typical of the nineteenth century

Almost hidden in a fine spray of water to subdue dust, the long needle nose of a
drill chatters into a hole in the rock marked with a blob of red paint. All along
the side of the stope a continuous line of red paint highlights a four-inch vein
of rock that, even to the uneducated eye, looks markedly different from the
rock below and above. It is a tightly packed bunch of white pebbles and
between them, here and there, a minute speck of gold gleams in the beam of
the miners' lamps.[1]

Gold mining in the rest of the world has not shown the same prosperity
in the twentieth century as in South Africa. Statistical tables show that
South Africa tends to produce ten times as much gold as her nearest rival,
Canada, and more gold than the rest of the world put together, except of
course Russia whose annual gold production is unknown. One of the problems facing the gold mining industry is that the price of gold on the inter-
national markets is fixed at $35 an ounce and it has remained as such since

[1] T. Green, *op. cit.*

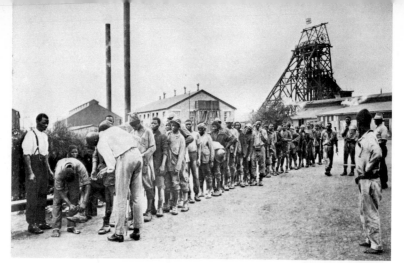

1934. The price is determined by the American Treasury which stands ready to buy and sell gold for dollars at this price. Costs of production have risen considerably since 1934 and consequently gold mining is not as profitable as it used to be. The decrease in profits has not had the serious effects that it might have done since much gold is a by-product in the mining of copper, lead, and zinc, and thus the cost of the gold output often bears little relation to the price. In addition, most mines operated chiefly for gold have one or more by-products that help to share the cost. In many mines silver is a by-product, and in some ores uranium is recovered as a by-product. In America 40 per cent of all gold produced is a by-product of the mining of base ores.

In America mines devoted entirely to gold mining are not very profitable because of the fixed price for gold on the open market plus the fact that the government refuses to subsidize the industry. Consequently gold mining has declined. The most profitable mine is Homestake Mine in South Dakota, producing one-third of the annual gold production of America. The industry received an enormous stimulus in 1965 when a very productive new mine was opened in the Tuscarora Mountains in north east Nevada. The Carlin Mine, as it was called, was the first new gold mine to be opened in America for over fifty years and it owed its existence to new technological skills. The largest flakes of gold were so small that they had to be magnified 1,800 times before they could be photographed. In spite of the minuteness of the gold particles, they occur so frequently in the rock that the mine yields one-third of an ounce for every ton of rock mined, as good as the yield from the Homestake Mine.

Gold mining in Canada has fared little better this century; rising production costs have not helped the industry. In 1966 gold mining on a commercial basis ceased in the Yukon. The Giant Yellowknife Mine on the northern shore of the Great Slave Lake is now Canada's largest mine; in Ontario two of the most productive mines of the 1960s have a very low life expectancy. The closure of uneconomic mines always raises problems for the mining community, as has been noted in the coal mining industry.

63 [opposite] Miners at a South African mine in the 1950s undergoing an examination of clothing at the end of a shift. 64 Miners relaxing in their compound at a South African mine. 65 Miners line up for their food at a South African mine

66   A gold dredger working in Colorado in a man-
made lake. Tin is mined in Malaya by a similar
method

These problems are particularly acute for some Canadian miners who can find little alternative employment in remote areas. The Giant Yellowknife Mine is a case in point since it is 700 miles north by air from Edmonton, the nearest large city. If the mine had to close, a community would probably die. Since 1965 the Canadian government has given subsidies to certain mines in order to prop up the less profitable. Australian gold mines face similar problems of increasing costs and declining profits. The Australian government has responded with subsidies but they are insufficient to encourage companies to carry out further exploration with a view to developing new mines. The goldfields of Western Australia centred on Coolgardie and Kalgoorlie are declining.

Russia is perhaps the most interesting country after South Africa as far

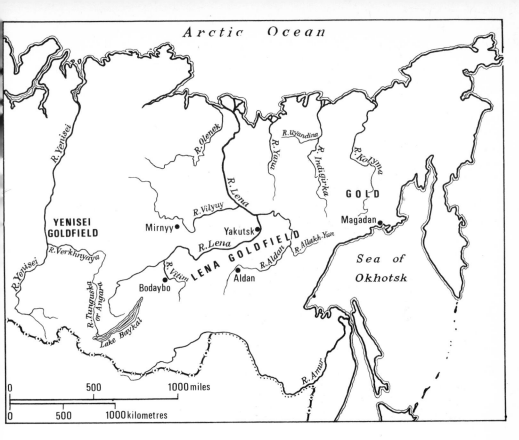

MAP NO. 5   GOLD MINING IN RUSSIA, LATE NINETEENTH AND TWENTIETH
CENTURIES

as gold mining is concerned. No one knows accurately the production
figures for Russian gold; some writers claim she lies second to South Africa,
but the simple fact is that since the mid-1930s no production figures have
been published by the Russians. Throughout the second half of the nineteenth century the Yenisey and Lena Goldfields Company had 25,000–
30,000 employees, many of them women. The gold lay near the surface in
gravels and was dug out by fairly primitive methods. The 1917 Revolution
caused a serious dislocation in the industry but under Stalin efforts were
made to develop known deposits and discover new ones. He initiated a
major expansion programme for Soviet gold in 1927, recruiting American
engineers working in Alaska to develop his mines since they were familiar
with the geological and climatic conditions of the far eastern provinces of

119

Russia. The principal recruit was John D. Littlepage who from 1928 to 1937 supervised the installation of machinery in alluvial goldfields and lode mines.

The Lena goldfield continued to be a major source of gold especially in the 1920s when production began at Aldan on the river Aldan, a tributary of the Lena, an area that had the richest deposit of alluvial gold ever discovered in Russia.

In 1931, east of the Lena, the major source of alluvial gold was discovered on the Kolyma river. It is estimated that three-quarters of Russian gold is mined here using primitive methods and a very large labour force. The labour force was alleged to be convict. Terence Armstrong in his book, *Russian Settlement in the North* quotes in a footnote the figures of one Paul Barton who published a book in 1957 in which he claimed a convict force of 3 million to 5 million in 1940 declining to 500,000 in 1953. Barton's figures are not substantiated. Other writers have quoted a convict force numbering 1 million by 1947. The whole area in which the Kolyma field lies was under the control of an organization called Dal'stroy from 1931 to 1957 and the produce of the goldfields is said to have paid for much of the early development of this almost virgin territory – towns, roads, ports, agriculture. It is interesting that the two most important gold-producing countries in the world, Russia and the South African Republic, owing their pre-eminence to a variety of not necessarily common factors, have in common a ready supply of cheap labour.

Much of Russian gold lies in alluvial sands and gravels and can be extracted by the fairly primitive method of digging the gravel out with pick and shovel. This technique was used in Lena field in the late nineteenth century and it was this backwardness that was one of the reasons for the recruitment of J. D. Littlepage in 1928. He came from the Yukon field, rich in alluvial gold, where considerable developments had taken place since the rush of 1898. As early as 1901 dredges were used in areas where the most readily available gold had been taken. Their method of working was described by T. A. Rickard, a distinguished American mining engineer who visited the Yukon:

> The barge is constructed at the bottom of a pit, excavated by the use of scrapers and horses, to a depth sufficiently below the expected water-level to ensure flotation and afford room for movement. Then the machinery is placed in position on the barge. As the water is admitted, the dredge floats, and when it starts its work it digs its own way, filling the pit behind as it advances.[2]

[2] T. A. Rickard, *Through the Yukon and Alaska*

Dredges similar to these were introduced into parts of the Lena fields. We cannot be certain if gold in the Kolyma fields was worked by dredges, a very large labour force suggests a more primitive method of working. Dredges are used in America today for alluvial gold mining based on the same principle as those introduced into Alaska and the Yukon in the early twentieth century.

What is the future for gold mining? No new deposits have been discovered in South Africa since the 1940s, production is declining in Australia, America, and Canada. Russia seems to be the only country with an expanding gold mining industry. It would appear that gold mining will never assume the importance it had in the nineteenth century nor will gold be obtained as easily. The lone prospector could make his fortune in the last century; these days are gone for ever since the most accessible gold deposits are exhausted. Gold can only be profitably mined by the large company prepared to spend enormous sums of money to recover gold buried deep beneath the earth's surface.

67　A small Russian gold mining village near Aldan, situated in the bleak countryside so typical of this most northerly part of Russia

9 Developments in iron ore mining since the mid-nineteenth century

DURING THE SECOND HALF of the nineteenth century important technological developments occurred in the iron industry. Demand for iron had been high throughout the century but processes developed by men such as Bessemer, Siemens, and Gilchrist made it possible to manufacture large quantities of steel cheaply, and the demand for iron ore continued to rise.

In Europe underground mining was the most common method of mining iron ore, but important developments took place in America. Until the 1890s progress was limited, miners used pick, shovel, dynamite, and drills to mine iron ore deposits which lay close to the surface. Abundant high grade ores were available principally on the northern and western shores of Lake Superior, and in the 1880s about 1 million tons of ore was being shipped from the Vermilion iron range on the north side of the lake to the steel town of Pittsburgh via the lake port of Duluth. Shortly afterwards even richer deposits were discovered to the west in the Mesabi range. They were near the surface, easily mined, non-phosphoric in content, and could be used in the Bessemer process for making steel. Andrew Carnegie, one of the creators of the giant United States Steel Corporation (founded 1901) recognized the enormous potential of the Bessemer process if applied to these ores. It was essential to mine large quantities of the ores and this was done by the introduction of steam shovels in the 1890s to strip off the overburden and remove the ore; a process already being developed in coal mining in Illinois. The introduction of machinery was vital once the most accessible ores were exhausted, since swift removal of ore was vital to the economies of the big companies involved. Large companies with enormous capital were created to develop the mines and the steel industry; the most famous being the United States Steel Corporation. Soon after its foundation this company owned and mined 65 per cent of Lake Superior ores, owned five large docks and an extensive fleet, ran its own trains on its own

68　A Bessemer converter. Developed in the 1860s, this
process made possible the cheap manufacture of steel

railway system, and produced 50 per cent of the total steel production of
America. The development of mining on a large scale using powerful
machinery was very much the work of the large corporation, a pattern that
was copied in many parts of the world where iron ore could be mined by the
open cast method.

From the 1890s the Mesabi range in the Lake Superior area was one of
the major sources of iron ore for the American steel industry and possibly
the mine where the most advanced techniques of open cast mining were
used. The overburden was no more than 65 feet in thickness and could be
removed without preliminary blasting. Enormously powerful steam shovels
were used to scoop the ore and dump it into railway wagons. Heavy demands
were made on the deposits and by the mid-twentieth century geologists were looking for alternative supplies. The most important deposits were
found in central Labrador in an area known as the 'Labrador–Quebec
trough', since they lie partly in the province of Quebec and partly in that of
Newfoundland. In 1954 mining began in a remote, uninhabited part of

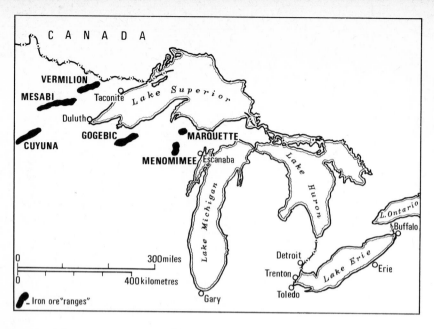

MAP NO. 6   THE LAKE SUPERIOR IRON ORE DEPOSITS

Newfoundland, 350 miles north of the St Lawrence town of Sept Isles, in an area known as Knob Lake. Six years later mines were opened up in a town called Gagnon in Quebec province about 100 miles north west of Sept Isles. In 1962 the third mining area in the Labrador–Quebec trough was opened up in the Wabush Lake area, some 150 miles north of Sept Isles. In all these areas the ore is mined from huge open pits by the open cast method; it has been estimated that the pit at Wabush Lake will be 3·5 miles long and 1 mile wide. The weather is very severe in these areas and some mining operations cannot be carried out in the coldest months from November to March. The crushing equipment will not work so only stripping of the overburden is done.

The need to mine in isolated areas of a country where few people live and communications are limited imposes special problems on the companies. They have had to construct railways through difficult country in order to move the ore to steelworks, and build towns in virgin territory for miners and their families.

Australia has enormous reserves of iron ore which have been intensively mined particularly since the end of the Second World War and their exploitation has posed in some areas some of the problems described in Canada.

124

69   An aerial view of Lac Jeannine mine, part of the
Quebec–Cartier iron ore project in Canada. Production rate
is about 20 million tons per year, and the picture shows the
40 foot terraces from which the ore is stripped

A good example of such a development is the mining area of Mount Tom
Price discovered in 1952 in the Hamersley Ranges of the northern part of
Western Australia. This is wild desolate country and until mining began
was almost uninhabited. Not until 1960 could a company be persuaded to
sink its money into the development of the deposits. In 1962 the Rio Tinto
Zinc Corporation, a multinational company with British, Japanese, and
American shareholders, opened up the area, having constructed a 176 mile
long railway to the mines and a town for the workers. Once the backing of
this corporation was enlisted it took only nineteen months to construct the
railway and begin production. The ore is mined by open cast methods; huge
excavators capable of removing 40 tons of ore at once are used. The area
has enormous reserves, but without the backing of a powerful company
mining would not have begun.

Underground mining of iron ore has not stopped in spite of the relative cheapness of open cast mining, since in some areas the overburden is too thick for it to be stripped off. In parts of America, for example Birmingham, Alabama, iron ore is mined underground, but the high cost of ore mined in this way has led to the steady decline of this type of mining.

Europe has for a long time mined considerable quantities of iron ore from underground mines. The rapid development of iron ore mining began in the late eighteenth century in Britain and in the early nineteenth century on the continent of Europe, largely because of the increased demands for iron goods. In Germany the development of iron ore mines was closely linked with the huge industrial combine owned by the Krupp family of Essen, whilst in France the Lorraine ore-field was developed, especially from the mid-century. In France the ore outcropped near the surface and the earliest mines were open pits, but as these deposits were worked out it was necessary to mine underground. Seams of up to 25 feet thick were discovered. These mines were worked by the methods discussed in Chapter 4 and benefited especially from developments in rock-boring tools.

In many areas in the late nineteenth century open cast mining was developed, almost certainly under the influence of developments in the American mining industry. In 1852 iron ore was discovered in Northamptonshire in England and since much of it lay near the surface it was easily dug out. As the century developed mechanical steam shovels were introduced to remove the overburden and ore. The industry never developed on the same lines as in America, but there were many small open cast mines employing small numbers of people; for example in 1933 in central Northamptonshire there were some sixty open cast mines scattered over a very small area, many of them employing no more than a dozen men. Since then more powerful machinery has been introduced into the open cast mines of northern Lincolnshire and in some cases as much as 100 feet of overburden is removed to reach the deposits – but the scale of the work is still small compared with American open cast mining.

Since open cast mining is usually cheaper than underground mining the former method is used whenever possible. Sometimes it is necessary to go underground because the costs of open cast mining are higher; the Kiruna deposits in Sweden are a good example. Mining began in the area in 1899 from open pits on the side of a mountain, but rising costs of terracing as the pits went deeper led the company to develop what is possibly the most modern underground mine in the world. The ore is loosened by drills and explosives and dumped by mechanical shovels down a series of chutes to a

70   The treatment plant for iron ore at Mount Tom
Price in the foreground, with the bleak terrain of the
Hamersley Ranges stretching out into the background

transport tunnel. Here a railway delivers it to automatic crushers and then
it goes by lifts to the concentrating mills. However, strip mining is carried
on in this area, too, in spite of the extreme temperatures of –40 °C. The
men working in this mine are supplied with fur coats as standard equipment, just as an underground miner is supplied with a reinforced helmet.

In spite of the harshness of the climate of this area within the Arctic
Circle and the problems created by the total darkness of the winter months,
efforts have been made to create an attractive community in an area of
swamp and stunted forest. In Kiruna there are skyscraper apartments and
comfortable bungalows, plus an indoor sports hall, supermarkets, and
cinemas. Miners have to be offered substantial benefits to work in such
areas. The contrast between towns such as this and those created by business tycoons of the late nineteenth century is striking. The towns of the
last century were isolated, devoid of decent living accommodation, without
elementary sanitation, and lacking in opportunities for recreation or medical attention. Some visitors to them were of the opinion that miners lived
like slaves. Towns like Kiruna might be isolated, but the need to attract

127

71   The iron ore loading complex cuts through Narvik
from the harbour up through the town. This
system is essential for the Kiruna ore from Sweden

men to work in remote areas has led to the building of towns with as many
amenities as possible, otherwise the miners ignore such areas for more
congenial jobs near large centres of population. Happily the day when
miners were at the mercy of their employers has long since disappeared,
something for which the growth of trade unions must take some credit.

The mining of iron ore by the open cast method now accounts for four-
fifths of the total world production and this trend will probably continue
because of the costs of underground mining. What is the future of iron ore
mining? Iron and steel are regarded as the most indispensable of modern
metals and are tremendously important in industrial countries. If all the
iron and steel now in use were suddenly to be removed, civilization as at
present understood would suffer a fundamental change. Happily there
seems little danger of this occurring since after aluminium, iron is the most
common metal of all on the earth's crust.

10 The mining of non-ferrous metals since the mid-nineteenth century

DURING THE NINETEENTH CENTURY world production of minerals such as copper, lead, and tin rose steadily with the increasing demand for them from countries that were rapidly becoming industrialized. New and more varied uses were found for such metals especially in the second half of the century. For example, demand for copper rose considerably because of its value to the electricity industry as an excellent conductor. Tin was not employed in industry on a large scale until the nineteenth century. In 1800 total world production was less than 9,000 long tons, by 1900 it had risen to 75,000 long tons, and by 1940 it had risen to 238,000 long tons. This rapid increase was due mainly to the extensive use of 'tin' cans as containers of meat, fish, and so on. Little of a 'tin' can is tin; the metal is used as a protective coat on a can of mild steel to prevent it from rusting.

The major mining area for these metals was no longer Europe, and this was particularly true of copper mining. As early as 1844 a company began to work the copper deposits of the Keweenaw peninsula of the Lake Superior region, and until the 1880s this area produced the major supplies in America. Then copper mining moved to the west and south west, to areas such as Nevada and Montana. These territories owed their early importance as mineral centres to their gold and silver deposits but when these were worked out their copper deposits were mined. For example, Butte copper mine in Montana first came into prominence as a gold and silver camp and was not mined successfully for copper until 1878. Once the isolation of these remote areas had been broken down, and the Indian tribes curbed (especially the Apaches), they became the leading copper-producing areas of America. The same type of development occurred in other parts of the world; Europeans opening up new territories exploited the minerals. In Chile most of the areas rich in minerals were discovered by the natives; the Spaniards paid most attention to the gold, silver, and tin,

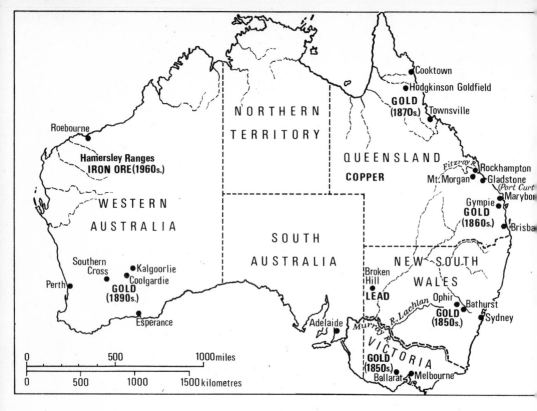

MAP NO. 7   SOME MINING AREAS IN AUSTRALIA

having abundant copper in Spain. Chilean copper first came into prominence when the British started to exploit the deposits in the 1870s, especially
those around Santiago, to be followed by the Americans. The development of copper mining in Ontario occurred when the Canadian Pacific
Railway reached Sudbury and the track layers cut through copper deposits.
This was the beginning of a mining industry in Ontario that produced
copper, nickel, gold, platinum, zinc, and silver, all from the same group of
mines in the Sudbury area. Base metals usually occur in very complex ores
and frequently more than one metal can be recovered from any one mine,
as this area has shown.

   Australia today is recognized as having enormous reserves of base metals.
From about 1850 onwards men were slowly becoming aware of the potential of the country and perhaps the most famous group of mines opened
were at Broken Hill in New South Wales. They were discovered in 1883 by
an immigrant German named Charles Rasp who was employed as a boun-

dary rider on a remote sheep station a few miles from Broken Hill. The first workings produced silver in quantity and by 1891 the largest company, the Broken Hill Proprietary, paid £1 million in dividends. However the miners discovered that there was lead and silver in the ore body and once a method was discovered to separate the lead and silver satisfactorily, lead was produced in considerable quantities. The new technique of ore separation was known as the 'flotation process'. Metallurgists found that in a mixture of crushed ore and liquid, certain minerals could be floated to the surface on bubbles of gas. This process made it possible to develop and extend the mines considerably and they became famous for the quality of the lead produced.

Descriptions of mining methods at Broken Hill in the late nineteenth century cannot be said to give a picture of underground mining in all ore mines, but some features were typical of many mines and of the developments which were taking place. The roof was supported by leaving pillars of rock which often contained valuable ore deposits, but in 1888 this method was replaced by the square set principle, for which the Comstock Lode was famous. The space between the floor and the roof of the stope was supported by a strong scaffolding of timber to give support to the roof and walls. Until about 1900 few of the mines used pneumatic drills, which was fairly

72 A Cornish tin mine in the late nineteenth century, showing miners climbing ladders to reach the stopes

131

unusual for metal mines. Holes were drilled by hitting a sharp mining steel with a hammer, one man hitting whilst the other gave the steel a quarter turn after each blow. Once holes were drilled they were filled with explosive and fired. The roof of the newly extended stope was timbered up if necessary and the ore was shovelled into trucks which carried it to the nearest connecting shaft or ore chute where it was discharged into trucks in the level below.

About the turn of the century new mining methods were introduced of which the horizontal cut and fill was the most popular. A long strip of ground was cut away over a period of many working days until the stope was like a large rectangular hall about 10 to 12 feet high, with only a few stacks of timber to support the roof. This was possible because the ore was being extracted from a more stable rock formation. Part of the bottom of the stope was filled with barren rock leaving a space of only about 5 feet between floor and roof. From this platform of barren rock, which enabled them to reach the ore, the miners drilled the roof and when they raised its height by removing ore from it they packed more rock beneath them. In cross section such stopes resembled a sandwich with a filling in the centre. This kind of stope was popularly called an 'underground open cut' by the miners.

By the time this mining method was used, large numbers of pneumatic drills were in use. As late as 1897 the largest mine rarely used pneumatic drills, but in 1900 it used sixty and piped air to them from a powerful compressor on the surface. These drills were not without their disadvantages. They were hard to erect or dismantle, for they weighed about 2 hundredweight each. They caused the workings to be filled with enormous clouds of dust which were a serious health hazard. Miner's disease, or pneumoconiosis, caught by breathing fine particles of dust caused by rock drills, was widespread until a new drill was introduced in the 1920s which had a hollow core through which water could be squirted continuously into the hole and settle the dust. Metal miners did not have the ever-present fear of

73 [*opposite*] Broken Hill in the 1900s, showing the square set system of timbering the roof. There was an ever constant danger of fire when such large quantities of timber were used.   74   Many coal and ore mines used horses underground to haul trucks along the main roadways; Broken Hill in the 1900s was no exception.   75   A photograph taken of a stope at Broken Hill in 1909; the wooden cylinder is an ore chute, down which ore was shovelled to the level below.

death from explosions caused by gases in the workings as in coal mines but their work was almost as dangerous. Miners in lead mines such as Broken Hill suffered serious illnesses from the lead dust they inhaled. Death from accidents such as falling down an ore chute, explosions of gelignite, and rock falls, were not uncommon plus the fact that familiarity with danger made men indifferent and carelessness caused death.

Since the beginning of this century the mining of non-ferrous metals has increased enormously to meet the needs of industry. New mining areas have been opened up and new methods developed. Zambia is such an area; it has been of some importance in copper mining since the 1920s and here a method of mining first developed in America has been applied with considerable success. Metal miners until this century used to follow the vein of ore carefully and when it became low grade they abandoned the level and attempted to find the high grade vein again. The new method, known as 'block caving', removed all the vein of ore whether low grade or high grade. Huge masses of rock are freed from the stope walls and undercut so that they fall into prepared cavities by gravity, being shattered in the process. The broken rock then slides down chutes into wagons on the level below where it is carried to the shaft. At the Creighton copper mine in the Sudbury basin in Canada the rock freed in the stope is allowed to slide from the sixth level (the highest level worked at the moment) to the fourteenth level where rough crushing is done. In modern ore mines the rock is loosened by charges of explosives, the holes are drilled deep by percussion drills. Modern drills are portable and stand on their own air legs. Some mines extend to a very considerable depth, for example the Creighton mine is now 8,000 feet deep and here the problems of ventilation, cooling, and dust control are very serious. Mining to this depth and using these techniques is common in many ore mining areas such as Chile, Zambia, and the Congo.

These non-ferrous ores are also mined by the open cast method. In 1892 open cast mining was begun at Broken Hill and as early as 1880 at the Bisbee copper mine in Arizona. The earliest open cast mines had no machines to dig the ore; they were worked by gangs of men renowned for their strength, who worked with pick and shovel. The twentieth century has seen an enormous extension of this type of mining and this has depended for its commercial success on two factors: the introduction of powerful machines capable of extracting large quantities of ore-bearing rock, and the continuous improvement of the methods of ore separation. Such is world demand for non-ferrous metals that it has become imperative for metallurgists

76 A miner using a drill with caterpillar tracks for ease of movement in a mechanized cut and fill stope at Broken Hill (1967).

to ensure that low grade ores can be used commercially.

Perhaps one of the most famous open cast mines in the world is the Bingham mine in Utah, America. Bingham Canyon is a narrow gorge and had been famous in the late nineteenth century for its gold, silver, and lead. Copper was known to exist but it was a poor grade, 1 to 1·5 per cent, and useless by the mining methods of around 1900. A mining engineer called D. E. Jackling insisted that copper mining could be made to pay if open cast mining was adopted and if a smelter could be built in the locality. Shortly before 1914 the mine began successful open cast mining and has continued ever since. The open pit began as a quarry on a hillside but such is the extent of the mine that the original hill has disappeared. There is now an oval hole 2 miles long, 1½ miles wide, and over ½ mile deep, with more than sixty wide terraces along which trains and trucks haul loads of blasted rock to the treatment plant. When Jackling offered his views for the profitable extraction of ore containing 1 to 1·5 per cent of copper he was laughed at; at Bingham today the ore contains only 0·75 per cent metal. At the mine they bulldoze 300,000 tons of rock every day and only by such large-scale operations coupled with the most sophisticated extraction techniques can the mine be made to pay.

77 Malayan tin mining: separating tin from gravel
swept by water into long sluices

Tin mining is perhaps the exception to the trend towards open cast
mining. Malaya produces more tin than any other country, using three
methods: underground mining, dredging, and hydraulic mining. The most
accessible deposits lie near the surface and these are mined first. In hydraulic mining a jet of water under heavy pressure is directed against a bank of
earth. The loosened gravel is swept by the water into a long sluice where the
cassiterite (oxide of tin), because of its high specific gravity, is easily separated from the gravel, which is lighter. This is the technique gold miners in
the nineteenth century used with the long tom. Dredging is done on a small
lake, sometimes artificially created, digging to a depth of up to 150 feet with
powerful dredges. The interesting feature of tin mining is that the bulk of it
is mined by dredging and hydraulicing but once the most accessible deposits are exhausted, mining moves to areas where the lodes are underground and fairly deep mines are developed.

So far the mining of metals known to man in antiquity has been discussed
but in the nineteenth century a number of metals unknown to man were
identified by scientists, and of these the most important was aluminium.

78 One section of huge open cast copper mine at Chuquicamata, Chile, showing the series of terraces cut out by power shovels which form working levels for further excavation

Not until the 1820s was it isolated, and a commercially successful reduction process was not devised until the 1860s, even then at a cost of £60 per ton. In 1886 a cheap method was devised by Héroult of France and C. W. Hall of America, working quite independently of each other. Since then an enormous number of uses has been found for aluminium because of its lightness and high conductivity of heat and electricity.

The only ore that yields aluminium readily and fairly cheaply is bauxite. Most deposits of bauxite tend to lie either on the surface or very near it and can be mined by open cast methods. In many ways the mining of bauxite differs little from the open cast mining of many metals but the finished product is typical of the link between mining and metallurgy.

The future of metal mining is worth considering. As the demand for metals increases, the metal mining industry relies increasingly on the metallurgist to devise methods of extraction for low grade ores. It seems obvious that mines will be driven to even deeper levels in order to supply the needs of the industries of the world. It is even more obvious that in the future careful reclamation of metals after use will be necessary in order

79   In contrast to illustration 78, miners tamping in powder
and placing wires before blasting with dynamite to
release ore far underground

to conserve world supplies. If the supply of a metal cannot keep pace with
demand, alternative metals and materials will be used. This has happened
in the case of lead, where there has been a very marked trend to substitute
other materials for it, such as aluminium for cable sheathing, copper for
water pipes, plastics for carrying corrosive liquids. Nevertheless, mining
still has a future, and it is very obvious that though the miner's tools may
have changed over the centuries, his value to society has only increased the
longer man has lived on the earth.

Principal Books Consulted

Man and Metals (2 volumes) by T. A. Rickard (McGraw-Hill)
Minerals in Industry by W. R. Jones (Penguin)
Roman Mines in Europe by O. Davies (Oxford University Press)
English Industries of the Middle Ages by L. F. Salzman (Clarendon Press)
De Re Metallica by G. Agricola, translated by H. C. and L. C. Hoover (Dover
    Publications)
The Rise of the British Coal Industry (2 volumes) by J. U. Nef (Frank Cass)
A Social History of Engineering by W. H. G. Armytage (Faber & Faber)
A History of Coal Mining in Britain by R. L. Galloway (David & Charles)
Annals of Coal Mining and the Coal Trade (2 volumes) by R. L. Galloway
    (Colliery Guardian Company)
Environmental Conditions in Coal Mines by J. Sinclair (Pitmans)
Coal Mines and Miners by Miles Tomalin (Methuen)
A History of Cornwall by F. E. Halliday (Duckworth)
The World of Gold by T. Green (Michael Joseph)
The American West by John A. Hawgood (Eyre & Spottiswoode)
A History of American Mining by T. A. Rickard (McGraw-Hill)
The Gold Rushes by W. P. Morrell (A. & C. Black)
The World's Great Copper Mines by B. Webster Smith (Hutchinson)
Early Diamond Days by Oswald Doughty (Longmans)
The Growth and Development of Australia by A. G. L. Shaw and H. D. Nicholson
    (Angus & Robertson)
The British Coal Trade by H. Stanley Jeavons (David and Charles)
Fundamentals of Economic Geography by W. Van Royen and Nels A. Bengtson
    (Constable)
Economic Geography by C. F. Jones and G. G. Darkenwald (Collier-Macmillan)
A History of South Africa by C. W. De Kiewet (Oxford University Press)
A Historical Geography of South Africa by N. C. Pollock and Swanzie Agnew
    (Longmans)
Case Studies in World Geography by Richard M. Highsmith Jnr (Prentice-Hall)
Russian Settlement in the North by Terence Armstrong (Cambridge University
    Press)
An Economic Geography of the United States since 1783 by Peter d'A. Jones
    (Routledge & Kegan Paul)
A History of Metals (2 volumes) by Leslie Aitchison (McDonald & Evans)

The Rise of Broken Hill by Geoffrey Blainey (Macmillan)

The Industrialization of Europe 1780–1914 by W. O. Henderson (Thames & Hudson)

A History of Technology by Charles Singer, E. J. Holmyard, A. R. Hall, and Trevor I. Williams (editors) (Clarendon Press)

The Economic History of the U.S.A.: Volume VII, The Decline of Laissez Faire by Harold U. Faulkener (Holt, Rinehart and Winston)

The Great Northern Coalfield, 1700–1900 by Frank Atkinson (University Tutorial Press)

Britain and Industrial Europe, 1750–1870 by W. O. Henderson (Leicester University Press)

Cambridge Economic History of Europe by M. M. Postan and H. J. Habakkuk *Volume II: Trade and Industry in the Middle Ages* (Cambridge University Press)

The Ancient World at Work by Claude Mossé (Chatto and Windus)

Mining Frontiers of the Far West, 1848–1880 by Rodman Wilson Paul (Holt, Rinehart and Winston)

The Growth of the American Republic (2 volumes) by S. E. Morison and H. S. Commager (Oxford University Press)

The New American Nation Series: The Far Western Frontier, 1830–60 by R. A. Billington (Hamish Hamilton)

The Economic Development of France and Germany, 1815–1914 by Sir John H. Clapham (Cambridge University Press)

Acknowledgements

The author and publishers wish to record their grateful thanks to copyright owners for the use of the illustrations listed below:

B. T. Batsford Ltd. for: 11
Broken Hill South for: 73, 74, 75
Camera Press for: 47, 51, 52, 59, 61, 67, 70, 71, 78
Charter Consolidated Ltd. for: 57
Crosby, Lockwood & Son for: 42
De Beers Consolidated Mines Ltd. for: 53, 54, 55, 56, 58
Mary Evans Picture Library for: map list tailpiece, 18, 20, 29
Federal Government Department of Information, Malaysia, for: 77
John R. Freeman & Co. Ltd. for: 16, 19, 24, 72
The Mansell Collection for: Introduction tailpiece, 15, 27, 32, 33, 34, 36, 37, 39, 40, 60, 62, 63, 64, 65, 66
McGraw-Hill Book Co. for: 7, 8, 9
Mining Magazine for: 12a, 13
The National Coal Board, London, for: 10, 17, 22, 23, 25, 45, 46
New South Wales Government for: 76
Royal Canadian Mounted Police for: 38
The Science Museum, London, for: jacket photo, 4, 12b, 12c, 14, 21, 41, 43a, 43b, 68
Stora Kopparbergs Bergslags Aktiebolag for: 26
The Trustees of the British Museum for: 1, 2, 3
The United States Information Service for: 35, 44, 48, 49, 79
The United States Steel Corporation for: 69
Mr Roger Worsley for: title-page, 5, 6, 28, 30, 31, 50

Index

Printed in Great Britain by Jarrold & Sons Limited, Norwich